Two-metre Antenna Handbook

Two-metre Antenna Handbook

F. C. JUDD G2BCX
FISTC, MIOA, Assoc.IPRE, A Inst. E

Newnes Technical Books

Newnes Technical Books
is an imprint of the Butterworth Group
which has principal offices in
London, Boston, Durban, Singapore, Sydney, Toronto, Wellington

First published 1980
Reprinted 1983

British Library Cataloguing in Publication Data

Judd, Frederick Charles
Two-metre antenna handbook.
1. Radio – Amateur's manuals
2. Radio frequency modulation – Receivers and reception 3. Radio – Antennas
I. Title
621.3841'352 TK9956 79-40024

ISBN 0-408-00402-9

Typeset by Butterworth Litho Preparation Department
Printed in Great Britain by Page Bros (Norwich) Ltd.

Preface

This book is intended for those using the now highly popular 2 metre band for the first time, although experienced operators on the VHF/UHF bands may find some of the new antennas such as the ZL series are worth trying out. I have not been able to include every type of antenna in a small book of this nature, or indeed specific antennas for the lower VHF band of 70 MHz or the UHF band of 430 MHz. The different antennas described herein should, however, provide a wide enough selection from which to choose at least one that will cater adequately for individual requirements and circumstances, particularly where space may be a problem. Any of the antennas illustrated can of course be scaled up or down for other bands and need only be re-dimensioned accordingly to produce the same performance.

I have also attempted to cover the 'basics' of propagation, transmission lines and matching to antennas, at least sufficient for a working knowledge, because these are subjects which, taken to the fullest extent, would have entailed writing a book three times as big. The reader will, however, find answers to what might be described as 'user problems' and particularly those associated with matching, feed cables and the often misunderstood function of VSWR (voltage standing wave ratio).

I have at one time or another used nearly all the antennas dealt with in Chapters 2 and 3, and have at least checked the performance of others with the UHF model antenna test equipment described in Chapter 5. The few of my own designs included, such as the Slim Jim omnidirectional antenna and the ZL beam, have already found favour with 2 metre operators in many countries. There are also details for two more antennas which have not hitherto been made known.

Many 2 metre operators have at one time or another assisted with tests on new antennas, especially G3JMU of Oulton Broad in Suffolk whose co-operation with daily 'long range' tests from my original QTH in London proved invaluable in the development of some of the antenna designs contained in this book. My thanks to them of course, to my wife Freda who undertook the typing of the original manuscript, and to numerous others who have supplied photographs and technical information.

One final point; since this book will also be read by radio amateurs worldwide, I have used the term *antenna* rather than *aerial* for the singular and *antennas*, not *antennae*, for the plural. Perhaps the more French *metre* rather than *meter* will be kindly accepted by readers of all nationalities when referring to wavelength.

Cantley, Norfolk

F. C. Judd
(G2BCX)

Contents

Chapter 1

Wave Propagation and Fundamentals

As it is common practice for radio amateurs to employ the same antenna for transmission and reception, it is important that a high standard of efficiency is maintained. A wrongly adjusted antenna, poor quality or wrong impedance feed cable, can all evoke high VSWR (voltage standing wave ratio), and consequent loss of radiated power and/or received signal and can have disastrous effects, as can a poor site for the antenna. Because in radio communication we are dealing with electromagnetic waves travelling through the earth's atmosphere, some knowledge of wave characteristics and how radio waves are influenced by propagation conditions is worth while. Such knowledge is an asset to successful DX on VHF for it is just as important to know when conditions are good as to have an antenna with top performance.

Wave characteristics

A radio wave is a combination of electric and magnetic fields with the energy divided equally between the two. If the waves could be originated at a point source in free space they would spread out in ever increasing spheres with the source at the centre. The speed at which the spheres expand is the same as the speed of light because light is also an electromagnetic wave. In empty space this is 299 792 077 metres or 186 282 miles per second. For general calculation however, these figures are rounded off to 300 000 000 metres or 186 000 miles per second.

In a remarkably short time a sphere growing outwards from a central point would be very large and to an observer on the spherical surface, if he could actually see the wave front in his vicinity, it would appear to be flat rather than round.

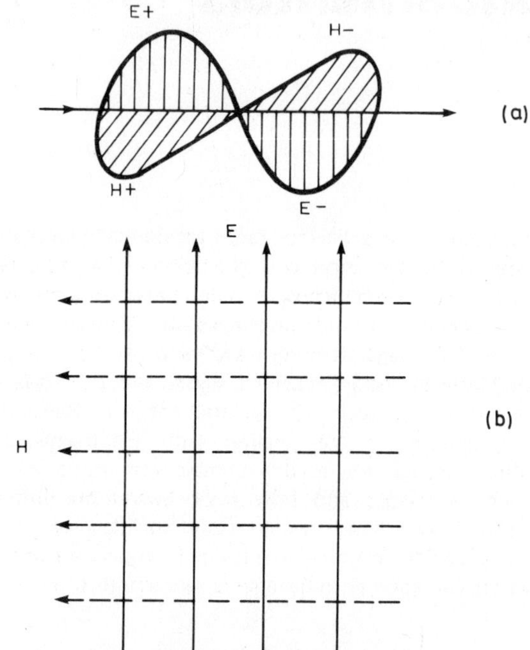

Fig. 1.1. (a) Radio wave configuration in free space. (b) Wave front at great distance from the transmitting antenna

A wave front that is far enough from the source to appear flat is called a *plane* wave, and radio waves always meet this condition, at least after they have travelled a short distance from the transmitting antenna. A typical representation of the lines of electric and magnetic force in a plane wave is shown in Fig. 1.1 in which (a) illustrates a section of an electromagnetic wave, moving in the direction of the long arrow. In the diagram, the lines, which should really be extended indefinitely, have been drawn in varying lengths to emphasise the idea of a wave. The electric field E is shown vertical and the wave in this case is vertically polarised. The magnetic field H is therefore horizontal, since the electric and magnetic fields are always at right angles to each other. However, the amplitude of the wave varies periodically (at the frequency of the transmission) from a maximum positive value and then through zero to a maximum negative value, i.e. the wave varies sinusoidally which is why antennas are tuned to resonance.

Wavelength

The section of wave illustrated in Fig. 1.1(a) is one wavelength long and its field is repeated cyclically along its path. If the wavelength were 1 metre, then 300 000 000 complete oscillations would pass any given point in one second. As the velocity of radio waves is 300 000 000 metres per second and the wavelength 1 metre, then the frequency (f) can be determined from

$$f = \frac{V \text{ (velocity)}}{\lambda \text{ (wavelength)}}$$

In this case it is 300 000 000 Hz or 300 MHz. An observer facing the wave would 'see' the field as shown in Fig. 1.1(b) which would be alternating at a frequency f. It is this alternating field that excites a receiving antenna and into which it imparts some of its power.

Wave reflection

In practice radio waves can be refracted and reflected, like light, by obstacles in the path such as hills, large buildings, metal structures – towers, gas holders etc. If the conductivity of such obstacles is good, much of the wave will be reflected and travel in other directions. If the conductivity is poor, some of the wave will be absorbed and a smaller amount reflected. Radio waves can of course travel through poorly conducting obstacles such as brick walls (albeit with some loss), and hence it is possible to use indoor antennas for transmission and reception. Conversely, highly conductive structures (like steel framed buildings) in the path of a radio wave can effectively prevent the wave from reaching a receiving antenna at all, or at least can reduce the strength very considerably.

Wave propagation

Radio waves travel in different ways according to the frequency in use and the polarisation of the wave as it leaves the transmitting antenna. For example, low frequency (long wavelength) vertically

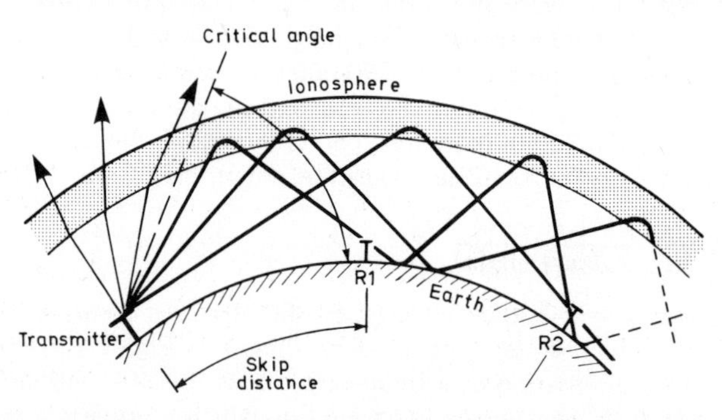

Fig. 1.2. How short wave (high frequency) radio waves are 'bounced' around the earth to distant points by reflection from an ionised layer in the earth's upper atmosphere. The angle at which waves are fed back depends on the angle of entry. Above the critical angle the waves are not bent enough to return to earth

polarised waves from a vertical transmitting antenna travel close to the surface of the earth, and this is commonly called *ground-wave* propagation. The distance for reliable ground-wave reception is limited because the wave gradually becomes absorbed by the surface over which it travels. However, at higher frequencies (shorter wavelengths) and with suitable transmitting antennas, radio waves travel upward at a relatively steep angle from the earth. They are then reflected back to earth at a distant point by an ionised layer in the earth's upper atmosphere called the ionosphere. This is known as *sky-wave* propagation, and as the frequency of transmission increases (wavelength becomes shorter), so greater and greater distances can be covered around the earth, as illustrated in Fig. 1.2. Thus the frequency of transmission and the antenna used at the transmitter can be selected to provide

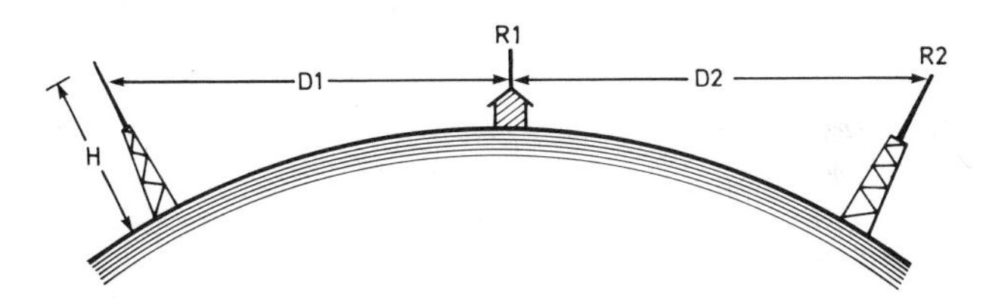

Fig. 1.3. Line of sight transmission and reception path at VHF and UHF

either short distance service (ground-wave propagation) using low and medium frequency bands, or long distance service (sky-wave propagation), using high frequencies. When the transmission frequency is very high (VHF), radio waves are no longer reflected by the ionosphere but pass right through it. Hence the transmitting antenna must be able to 'see' the receiving antenna, and the propagation is by *'space wave'* (more commonly known as 'line of sight'). Provided there are no large obstacles such as hills and high buildings in the path, the range can be to the horizon or further, depending on the height of both the transmitting and receiving antennas.

The structure of the atmosphere near the earth's surface is such that under normal conditions the waves are bent into a curved path that keeps them nearer to the earth than a true straight line would. This effect can be approximated mathematically by assuming that the waves travel in straight lines but that the earth's radius *is increased* by one third. On this assumption the distance from the transmitting antenna to a receiving antenna at ground level on the horizon is given by:

$$D \text{ (km)} = 3.93 \times \sqrt{H} \text{ (metres)}$$

or $$D \text{ (miles)} = 1.415 \times \sqrt{H} \text{ (feet)}$$

where H is the height of the transmitting antenna as shown in Fig. 1.3. This assumes that the earth is perfectly smooth and that there are no obstacles (hills etc.) in the path of the radio wave. If reception at greater than horizon distance is required then the height of the receiving antenna must be increased as at R2. The maximum 'line of sight' range is now $D1 + D2$.

Tropospheric propagation

Weather conditions in the atmosphere, at heights from a few hundred metres to a kilometre or so, are at times responsible for bending VHF radio waves downward. This tropospheric refraction makes VHF communication possible over greater distances than can be achieved with normal line of sight propagation. The amount of bending increases with frequency and transmission and reception improve as the frequency goes up from about 50 MHz. Hence long distance reception of television and VHF/FM broadcast stations is sometimes possible. Refraction in the troposphere takes place when masses of air become stratified into regions having differing dielectric constants. If the boundary between the two masses of air is sharply defined, reflection as well as refraction may take place for waves striking the boundary at grazing angles. The most common cause of tropospheric refraction is temperature inversion. Normally the temperature of the lower atmosphere decreases at a constant rate of approximately 2°C per 300 m (3°F per 1 000 ft) of height. When this rate is decreased for some reason, a temperature inversion is said to take place and greater than normal wave bending occurs. Because the atmospheric conditions that provide tropospheric refraction seldom last for any considerable period of time, the strength of received signals usually varies, i.e. fading occurs. A tropospheric wave also generally maintains its polarisation throughout its travel.

Atmospheric ducts

In some parts of the world, particularly in the tropics, or over large areas of water, temperature inversions are present almost continuously at heights of a few hundred metres or less. The boundary of the inversion is usually well enough defined so that waves travelling horizontally are 'trapped' by the refracting layer of air, and continually bent back to earth. The air layer and the earth form the upper and lower walls of a duct in which the waves are guided, in much the same way as microwave transmission in metal waveguides. The waves therefore follow the curvature of the earth for distances far beyond the horizon, sometimes for hundreds of kilometres.

Line of sight propagation

The higher one gets the further it is possible to see, and this simple fact applies very much to VHF and UHF radio propagation over line of sight distances. However, the radio amateur is usually obliged to erect antennas at fairly restricted heights (restricted that is in comparison with the high masts on elevated sites used by broadcast stations). Fortunate indeed are those amateurs who

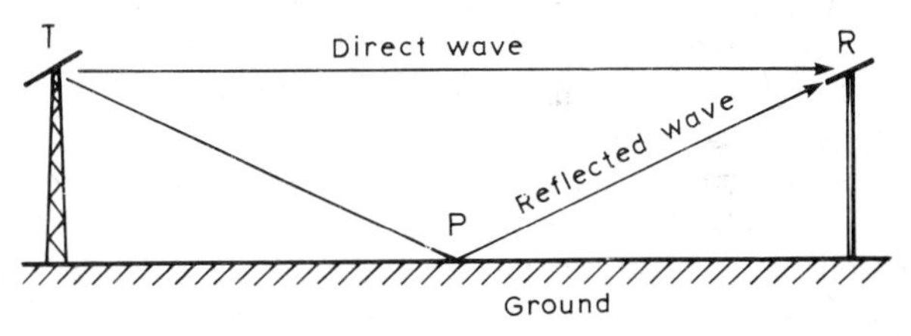

Fig. 1.4. Direct v reflected wave from ground. Depending on path difference the waves can arrive in or out of phase at the receiving antenna

live on high ground. Wave propagation at VHF and UHF will normally be over a path parallel to ground and therefore of limited horizontal distance. As Fig. 1.3 illustrates, this distance can be increased by greater elevation of both the transmitting and receiving antennas, but one problem remains which Fig. 1.4 may serve to clarify. Whilst a wave leaving the transmitting antenna

T will travel straight to the receiving antenna R, part of it will also travel to the ground and be reflected at some point P. Depending on the path length difference, the reflected portion of the wave could arrive at R either in or out of phase with that travelling direct. The result could be a strong signal due to the waves arriving in phase, or a weak signal due to the waves arriving out of phase (because in-phase signals add up and out-of-phase signals cancel each other). It frequently happens that movement

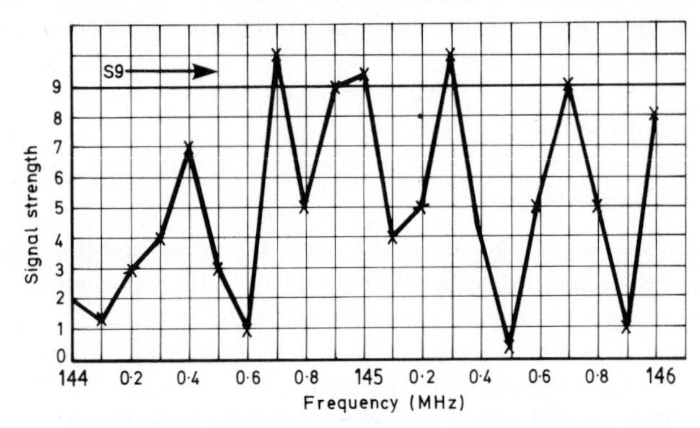

Fig. 1.5. Variation in signal level over a constant path at various frequencies across the 2 metre band. Transmitting and receiving antennas both vertical

of a transmitting or receiving antenna by a fraction of a wavelength in one direction or another, or a change in frequency of operation, will bring an otherwise weak signal up to full strength. This effect is particularly noticeable when vertically polarised waves are used, i.e. vertical transmitting and receiving antennas are in use. Fig. 1.5 is from actual measurements across the 2 metre band (144 to 146 MHz), and shows the quite violent fluctuation in signal strength versus frequency using vertically polarised antennas and transmitting over a line of sight path.

In practice VHF and UHF wave propagation is much more seriously affected by shadowing due to high buildings, heavily wooded ground and hills and by diffraction and absorption loss. The higher the frequency the more serious this becomes, and because of the large number of these variables in the line of propagation it almost impossible to calculate working ranges and signal levels at the receiving end, except for a more or less true line of sight path. Really long distance operation on 2 metres and 70 cm (145 MHz and 420 MHz) is therefore almost entirely

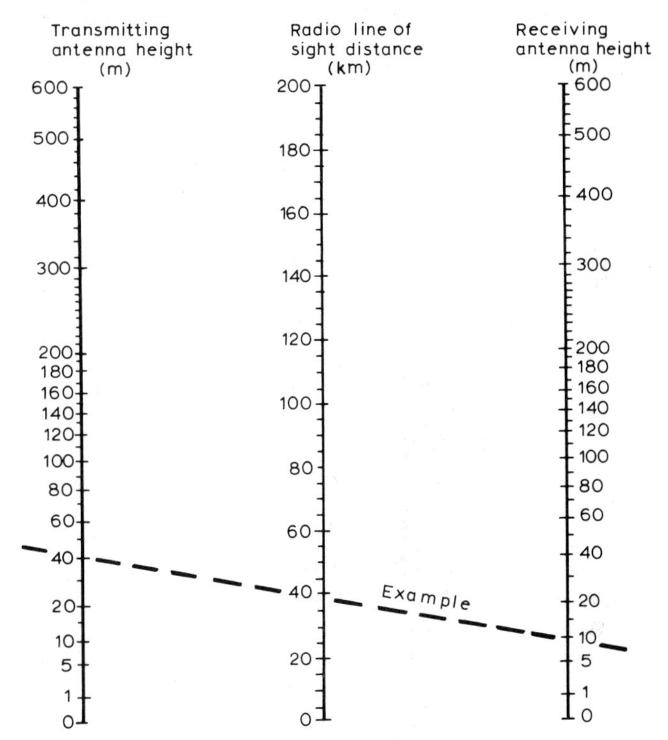

Fig. 1.6. Radio line of sight distances for different heights of transmitting and receiving antennas. Example shows how to obtain distance

dependent on tropospheric lift, or, as radio amateurs refer to it, good lift conditions. The circumstances required to set up an antenna system and confidently expect to work some given distance at any time and regardless of 'lift conditions' of one kind or another are indeed rare.

The most valuable aid to successful long distance operation is ample height of both transmitting and receiving antennas, and relatively low angle (parallel to ground) radiation.

Fig. 1.6 gives some idea of line of sight distances for various transmitting and receiving antenna heights. Typical power/range

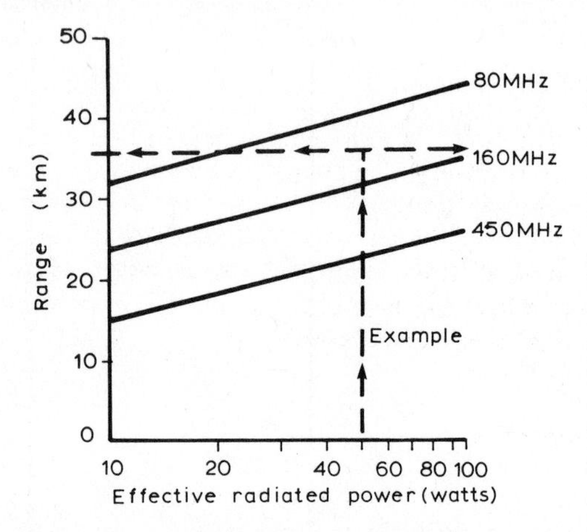

Fig. 1.7. Probable range for given ERP at VHF for a base station using an omnidirectional unity gain antenna transmitting to mobiles

graphs, for mobile working with a base transmitting station using a unity gain omnidirectional antenna at a height of 30 metres, are given in Fig. 1.7.

Other propagation conditions

The reading list given at the ends of chapters will provide much more detail on other propagation conditions than can be included here. However, some other media of propagation at VHF and UHF (tropospheric scatter, ionospheric scatter, auroral reflection, meteorite scatter and sporadic E) will be discussed here.

Tropospheric scatter

This is very similar to normal tropospheric propagation except that return signals are reflected or refracted by cloud banks along the transmission path. Attenuation is greater so higher power may be required but distances of 800 km (500 miles) or so are possible.

Ionosphere scatter

Ionospheric clouds provide a reflecting surface for low angle radiation of relatively high frequency waves and by a series of hops a wave may travel to a distant point over a more or less straight path. This is usually known as forward scatter and, although it is a fairly regular occurrence, high power may be required to offset the high attenuation. It is more applicable to the 21 and 28 MHz amateur bands.

Auroral reflection

This occurs from zones around the polar regions and during ionospheric or magnetic storms when there is a marked auroral activity. Frequencies at which reflection from this medium takes place are up to about 150 MHz, and propagation is characterised by very rapid flutter which makes telephony difficult to read but does not greatly effect CW. Radiation must obviously be directed toward the polar caps, north or south accordingly, and the receiving antenna must be so directed as well.

Meteor trail scatter

This is a short duration phenomenon often lasting less than a minute. It is produced when a meteor passes through the upper atmosphere leaving a trail of ionisation. This will cause partial reflection of VHF waves. Frequencies from around 21 MHz and up to 145 MHz can be used. High power and very efficient antennas are essential.

Sporadic E

At about the same height as the normal E layer, highly ionised clouds are often formed sporadically and at random. They vary in intensity and move rapidly from south east to north west in the northern hemisphere. They occur mostly between May and August and sometimes in mid-winter but the seasonal appearances are reversed in the southern hemisphere. The MUF (maximum usable frequency) is a function of ionisation density and generally at around 50 MHz. DX is often possible on 2 metres but at a rather low percentage compared with, say, 50 MHz. Distances of 3 000 km (2 000 miles) or more are possible with the right conditions.

Field strength

In itself this is not a very predictable parameter at VHF and UHF except for relatively uninterrupted line of sight distances. Calculated and measured field strengths show reasonable similarity at 145 MHz for given types of transmitting antennas at various heights above ground (assuming zero feet above sea level) and for a transmission path over relatively flat, i.e. not mountainous, country. Fig. 1.8 shows the calculated and measured average received signals obtained using a transmitting antenna at a height of 20 m (60 ft) with an ERP (effected radiated power) of around 100 W and operating at 145 MHz. Signal levels plotted are within the broad shaded area and where reception was from reasonably efficient antennas at a mean height of 10 m (30 ft).

The measurement of field strength is difficult to achieve accurately and the radio amateur will almost certainly have to set his own reference to obtain useful information. Signal strength meters on commercially made receivers and/or transmitter-receivers are virtually useless except for comparison purposes. Much the same applies to the measurement of field strength for determining the gain of transmitting antennas.

Polarisation

The polarisation of a half-wave antenna is always the same as the direction of its axis when the distance is great enough for radiation to be considered as a plane wave. This means that the electric

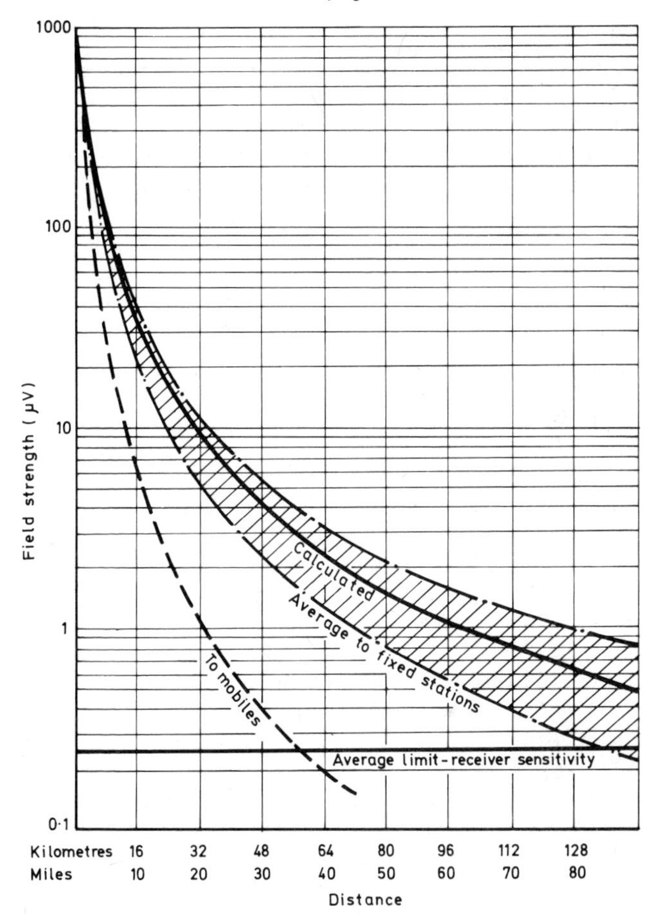

Fig. 1.8. Calculated v measured (shaded area) field strength. Transmitting antenna 20 m (60 ft) above ground. Receiving antennas average 10 m (30 ft) above ground

field is the same as the direction of the antenna (vertical antennas therefore radiate vertically polarised waves, and horizontal antennas, horizontally polarised waves as in Fig. 1.9). With antennas having a number of half-wave elements all lying in the same direction, the polarisation will be in that direction. If the antenna system consists of both vertical and horizontal elements radiating *in-phase,* the polarisation will be *linear,* i.e. tilted between vertical and horizontal (sometimes referred to as slant polarisation). If vertical and horizontal elements are fed *out of phase,* the polarisation will be either circular or elliptical. With circular

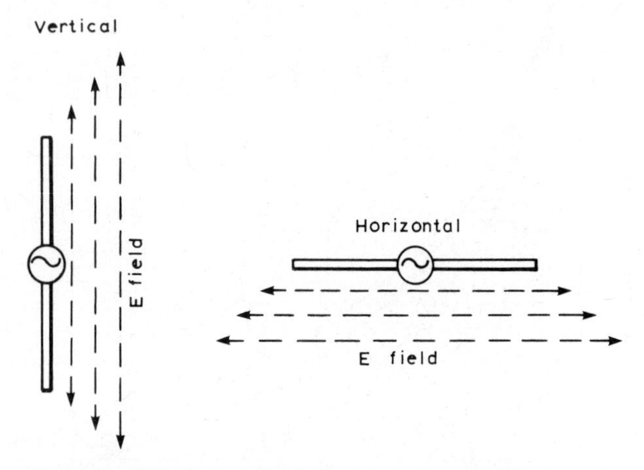

Fig. 1.9. Polarisation. Vertical antennas for vertical polarisation and vice versa

polarisation the wave front passing the receiving antenna rotates every quarter cycle between vertical and horizontal thus making a complete rotation every 360°. Circularly polarised antennas are frequently used for space communication.

The polarisation of a wave can however become changed along the path of travel and this is quite common at VHF. It is not very serious, at least not normally serious enough to warrant vertical to horizontal or vice versa rotation of the antenna. Tests carried out on both near and distant stations using vertically polarised transmitting antennas, e.g. distant repeater stations, revealed some orientation of polarisation, as Fig. 1.10 illustrates.

What is more important is the effect of polarisation when working long distances. Investigation and tests over a number of years have revealed, for 2 metres and higher frequency bands, that horizontal polarisation is far superior and much less subject to path attenuation and fading under tropospheric and similar lift conditions. Over a parallel to ground path vertically polarised signals suffer a very high rate of attenuation and are much more

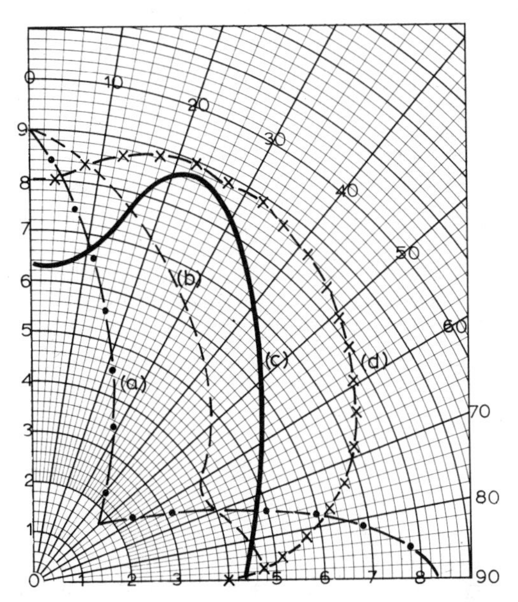

Fig. 1.10. Effect of polarisation change over transmission path. (a) From nearby fixed station using ground plane. (b) From nearby fixed station using free space vertical. (c), (d) From distant repeater stations. Slight change in polarisation by receiving antenna shows increase in signal strength particularly for (c)

prone to phase effect. Vertical polarisation has only become more widely used because of mobile/repeater operation and therefore the use of omnidirectional (vertical) antennas. The majority of 2 metre operators miss out on many DX contacts by not using horizontally polarised antennas.

Resonance

Antennas are normally constructed from metal of good conductivity e.g. copper, aluminium etc., but to be efficient they must be resonant which means they must be electrically tuned to the frequency of transmission (or reception). The shortest length of conductor that will resonate to a given frequency is one just long enough to permit an electric charge to travel from one end to the other and back again in the time for one cycle of the wave. If, as already explained, the speed at which the charge travels is equal to the velocity of light (300 000 000 metres per second), the distance it will have to cover in one cycle will be one wavelength. Because the charge traverses the conductor *twice*, the length needed to allow the charge to travel a whole wavelength will be $\lambda/2$, or one half-wavelength (in practice slightly less however, because of the velocity factor of the conductor). Radio frequency currents travel more slowly in metal conductors than in free space. The current and voltage distribution in a half-wave antenna (or dipole) is shown in Fig. 1.11. When a charge reaches

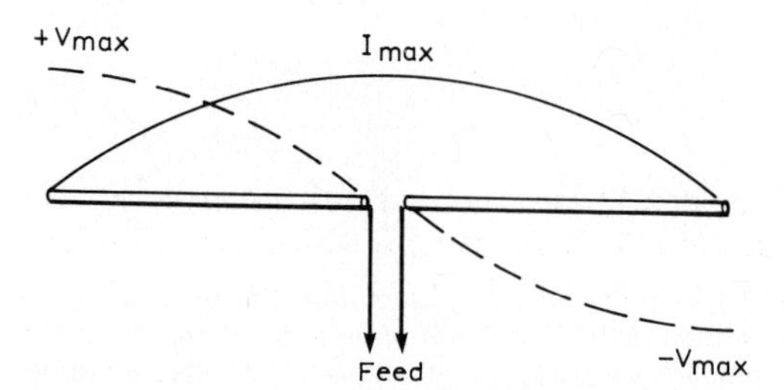

Fig. 1.11. Current and voltage distribution on a half-wave dipole

the end of the dipole its direction and phase are reversed, so an alternating flow of current is achieved to the full cycle although current and voltage are themselves always in phase opposition, i.e. when voltage is maximum, current is minimum and vice versa. Provided resonance is maintained, the current maximum will be at the centre of the dipole and at this point the impedance will be low, about 72 Ω, which makes it suitable for direct connection to 70/75 Ω coaxial cable or open wire feeder of the same impedance. The physical length for a half-wave dipole at resonance is less than the electrical length because of the velocity factor of the conductor. The equations

$$\text{length in metres} = \frac{150 \times K}{f} \quad (f = \text{frequency in megahertz})$$

$$\text{length in feet} = \frac{492 \times K}{f}$$

$$\text{length in inches} = \frac{5905 \times K}{f}$$

give K the velocity factor of the conductor used to make the antenna which for practical purposes can be taken as 0.95. The length of a dipole for the 2 metre band (145 MHz centre frequency) would be

$$\frac{150 \times 0.95}{145} = 0.98 \text{ m (38.6 inches)}$$

Directivity

The intensity of radiation from an antenna and/or its ability to receive signals is never the same in all directions and may indeed be zero in certain directions. Although no antenna exists that will transmit or receive equally in every direction around it, it is convenient to assume one that does. This hypothetical antenna is called an *isotropic antenna*, and is frequently used as a standard for comparing the performance of other antennas. The radiation or pickup pattern of an isotropic antenna would in fact be a sphere and the antenna itself considered as a point source. By comparison the radiation pattern of a half-wave antenna is, if taken in all directions around it, doughnut shaped as shown in Fig. 1.12(a). This pattern remains the same whether the antenna is vertical or horizontal provided it is in free space or, in practice, at least several wavelengths above ground. In Figure 1.12(b) the doughnut shaped pattern has been cut in half so we can see the 'flat' pattern broadside to the antenna which remains the same regardless of whether it is vertical, as in the diagram, or horizontal. The directivity is therefore, broadside in two directions (bidirectional) and with no response from the ends. Looking along the

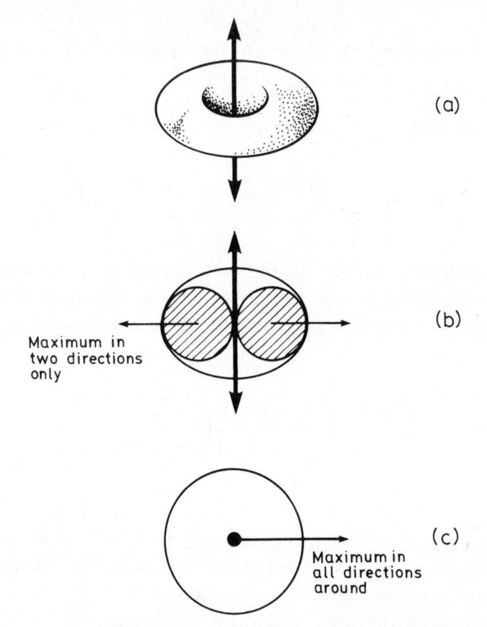

Fig. 1.12. (a) Viewed in three dimensions the radiation field from a dipole is 'doughnut' shaped. (b) With the doughnut cut in half the radiation field broadside to the antenna is a figure of eight shape. (c) Looking at the end of the antenna, radiation is equal in all directions around it

length of the antenna however, from either end, as in Fig. 1.12(c), the pattern becomes a complete circle (looking down on the doughnut) so the directivity is the same in all directions or, to use the correct term, omnidirectional. The usual field pattern for

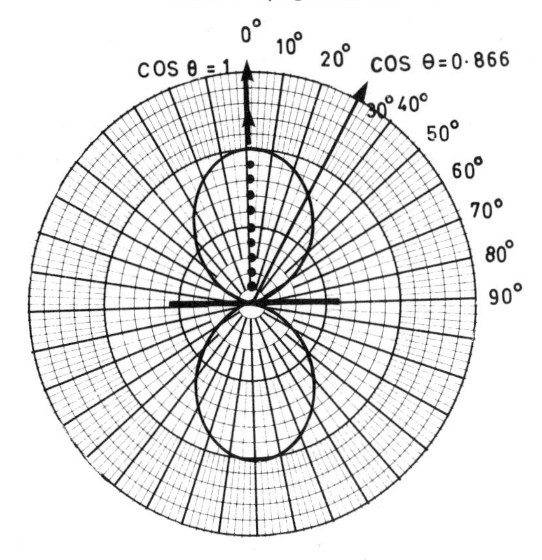

Fig. 1.13. Radiation at different angles from a dipole broadside to the element is approximately cos θ when 1 = maximum radiation at 0 degrees

a horizontal dipole is shown in Fig. 1.13 where the arrows indicate maximum directivity. This pattern is sometimes called a cosine pattern because it can be plotted from cos θ = E (intensity at various angles) when E_{max} (at zero degrees) = 1.

Gain

Antennas with directive properties, and these include the dipole, are said to have gain, which for measurement purposes is sometimes related to the hypothetical isotropic antenna which is assumed to have no gain because it radiates equally in all directions. By comparison, the gain of a dipole at maximum directivity, as in Fig. 1.13, is about 1.64 dB, or more correctly, 1.64 dBi, the 'i' indicating gain over an isotropic. However, the dipole is also accepted as a standard antenna and its gain is therefore referred to as 0 dB. Most manufacturers use this reference but if they wish to make an antenna performance look better, they quote gain with respect to an isotropic. For example, a beam antenna may be said to have a gain of 6 dB over a dipole or 7.64 dB over an isotropic, which should be clearly stated as 6 dBd (d for dipole) or 7.64 dBi respectively. Gain in an antenna system is only obtained by decreasing the arc or arcs of directivity which is

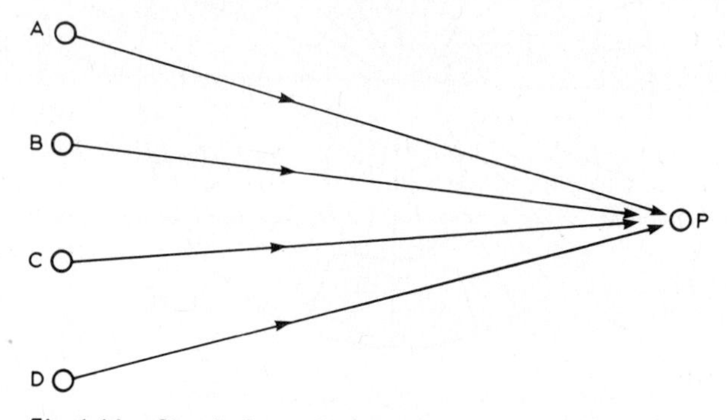

Fig. 1.14. Signals from all the radiators arrive in phase at the receiving point P so the signal level at this point is four times (or 6 dB) greater

accomplished by increasing the number of active elements such as dipoles, in which case the antenna may be called an active array. Gain, which should really be called directivity gain, can also be obtained by employing a reflector and/or a number of director elements in conjunction with one or more active radiators; usually dipoles. These are known as parasitic arrays as the reflector and directors are not normally active, i.e. they are not directly driven with r.f. power.

Directive arrays

One method of obtaining gain and directivity is to combine the radiation from a number of half-wave dipoles, and Fig. 1.14 may help to clarify the idea. Assume that the current in dipole A produces a certain level of signal E at some distant point P. The same current in any of the other dipoles will also produce the same signal level at P. If for example dipoles A and B only are operating each with a current I, the signal at P will be $2E$. With A, B and C operating the field will be $3E$ and with all the dipoles operating will be $4E$. As the signal received at P is proportional to the square of the field strength, the relative signal received at P is 1, 4, 9 and 16, depending on whether one, two, three or four dipoles are operating. The power gain is directly proportional to the number of elements used and where one element would yield a power gain of only 1 i.e. 0 dB. Four elements would produce a power gain of 4, which is 6 dB. It is important to remember however, that the fields from each separate dipole must be in phase at the receiving point and the currents in each must be identical. The elements must be separated sufficiently to reduce coupling between them which would otherwise affect radiation resistance etc.

Parasitic arrays

These are multi-element arrays containing at least one active or driven element, usually a dipole, and a reflector and a number of

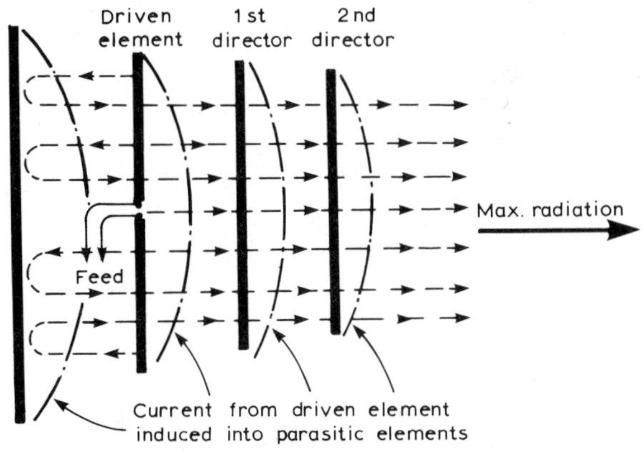

Fig. 1.15. The principle of a parasitic array

directors which are not driven and therefore classified as parasitic elements. A parasitic element obtains its power through electromagnetic coupling with a driven element (see Fig. 1.15). One of the most commonly used parasitic arrays is the 'Yagi' named after its inventor. This is popular because even a large array is relatively easy to construct and match to its feeder. Yagi antennas can be constructed with as few as three elements or as many as 12, to produce a gain over a dipole of 14 dB, 15 dB, or more. A pair of arrays may also be stacked one above the other, or bayed side by side, to produce even more gain. Generally this extra gain is relatively low, in the region of 3 dB greater than that obtained with a single array.

Radiation patterns

These are graphs showing the directivity gain of an antenna at all angles through 360° horizontal and vertical, plotted in Cartesian or polar co-ordinate form. They are valuable to radio operators as

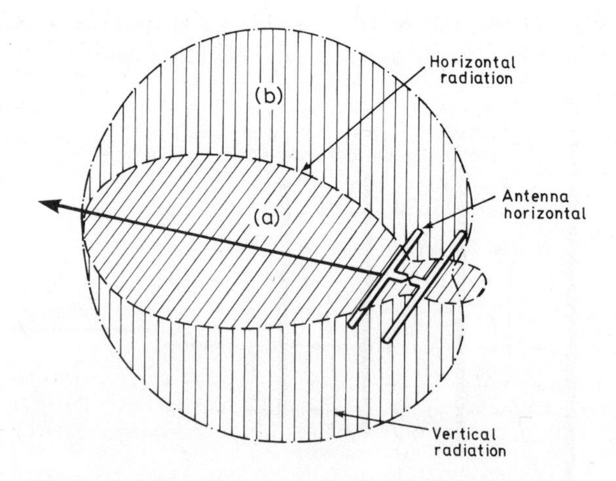

Fig. 1.16. The horizontal radiation pattern is not always the same as the vertical pattern, as in the case of this 2-element beam antenna, the ZL special (see Fig. 5.10)

they tell the complete story about the coverage as well as the gain of an antenna. Typical radiation patterns for a small beam antenna are given in Fig. 1.16 in which (a) is the plot for the horizontal field over 360°. It is apparent from this why it is necessary to be able physically to rotate antennas of this nature in order to get maximum power in a required direction. Pattern (b) shows the vertical angle radiation.

Antenna size

It is not generally realised that the physical size of an antenna is an important factor in performance. In reception the larger the area of antenna presented to the incoming wave front, the greater will be the strength of signal received. For example an array for say 430 MHz would have to present the same area to the wave front in order to receive the same level of signal as would be obtained from a similar array operating at 145 MHz. In other words the 430 MHz antenna would have to be physically larger, which means of course having a stacked and/or bayed system. This is rather like saying that a small bucket can be filled with water and so also can a large bucket but the smaller one holds less water. Although in the example given the two arrays may have the same gain, the physically larger, 145 MHz antenna, gathers more signal.

Reading list

Subject	*Reference*
Wave propagation and antenna fundamentals	*ARRL Antenna Book*, American Radio Relay League, available in UK from Radio Society of Great Britain, 35 Doughty Street, London, WC1N 2AE.
As above	*Radio, TV and Audio Technical Reference Book*, S. W. Amos, Newnes-Butterworths, 1977.

Propagation VHF and UHF	*Radio Wave Propagation VHF and Above*, P. A. Matthews, Chapman and Hall.
Notes for newcomers on antennas	*A Guide to Ameteur Radio,* 17th ed., Pat Hawker, G3VA, Newnes-Butterworths, 1979.
Antenna knowledge requirements for UK Radio Amateurs Examination	*The Radio Amateurs Examination Manual* and *Radio Amateurs Examination Questions and Answers.* Both published by R.S.G.B., 35 Doughty Street, London WC1N 2AE.

Chapter 2

Omnidirectional Antennas

Omnidirectional antennas have gained considerable popularity with the growth of mobile and repeater operation, mainly because a repeater station has of necessity to use an antenna that radiates equally in all directions, and because vertical antennas are more amenable for mobile operation. The half-wave dipole is the simplest of all antennas but is not necessarily the easiest to feed and use at VHF. The feed point impedance is nominally 72 Ω and if the antenna is to operate efficiently it requires a balanced feed. Most 2 metre transceivers have the now almost universal antenna output/input impedance of 50 Ω. This rules the dipole out unless provision is made for (a) a feed impedance step up and (b) a balanced to unbalanced feed. Also, as a dipole is centre fed, the feed cable has to run at right angles away from the antenna before dropping down to the transmitter otherwise it can cause severe interaction and reduce working efficiency. However, there are simpler ways of feeding a dipole that are just as efficient, especially one that is vertical.

The J matched half-wave

The method is to 'end feed' as in the arrangement shown in Fig. 2.1, which is known as the J match. Such an antenna may be made of stiff wire (even galvanised iron wire) or thin tube (better) virtually in one piece and may be supported at its base at the top of a mast. The feed cable may be 50 Ω coaxial and this can come away straight down from the antenna. A balanced feed is not important if the method of connection indicated here is used. All that is necessary to achieve maximum power feed to the antenna and

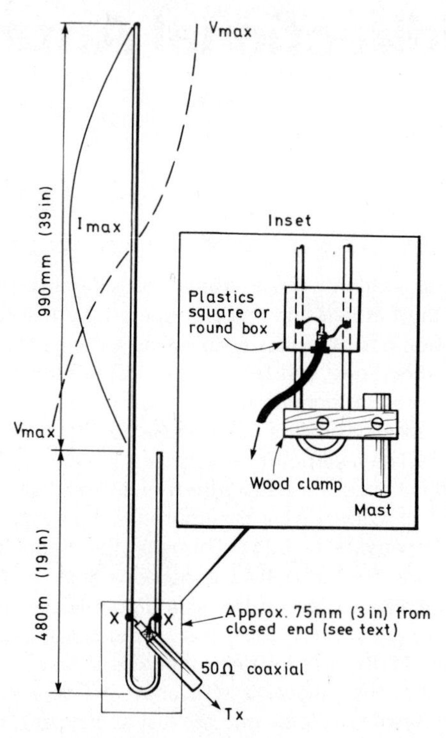

Fig. 2.1. Construction of a simple J matched end-fed half-wave dipole for omnidirectional operation

minimum VSWR (voltage standing wave ratio) in the feed cable, is to adjust the feed points (X) both up or down and then finally solder to the stub elements. A small plastics box around the feed point suitably sealed against the entry of rainwater is essential, but do take the cable right into the box to prevent water getting into it also. Rainwater in coaxial cable will ruin it.

This J match antenna operates on the principle that the quarter stub provides a high impedance feed to the half-wave section, i.e. a voltage feed. From the top or open end of the stub the impedance falls from high to low moving toward the closed or bottom end at which point it becomes zero. Although this antenna is vertically polarised, since it is operated vertically, it is omnidirectional as well, but as with all end-fed vertical dipoles the radiation at vertical angles to ground tends to be tilted upward, i.e. toward the end of the antenna.

The ideal is to have vertical angle radiation virtually parallel to ground – straight out from the antenna – and this is what is achieved with a simple to make antenna designed by the author, namely 'The Slim Jim'.

The Slim Jim

This is a vertically polarised omnidirectional free space antenna for 2 metres, but it will operate in the same way for other higher, or lower, amateur bands by scaling the dimensions accordingly. It has a radiation efficiency 50% better than a ground-plane antenna due to its low angle radiation, it is unobtrusive and has no ground plane radials and therefore low wind resistance. Why

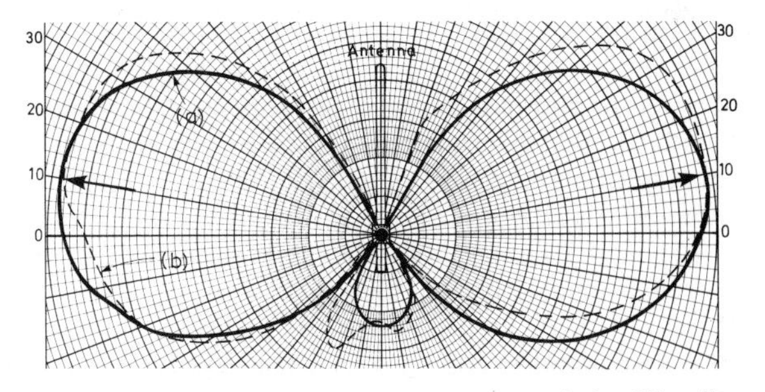

Fig. 2.2. Vertical angle radiation pattern of the Slim Jim antenna. (a) Theoretical. (b) Plotted from a 2 metre prototype

Slim Jim? Well this stems from its slender construction (it is less than 1.5 m (60 in) long for 2 metre operation) and the use of a J type matching stub (J integrated match = JIM) that facilitates feeding the antenna at the base thus overcoming any problem of interaction between feeder and antenna. The feed impedance is 50 Ω.

Why is the Slim Jim so much more efficient than the popular $^5/_8\,\lambda$ or other ground planes, despite the latter's claimed 3 dB gain over a dipole? The polar diagram in Fig. 2.2 provides the answer. The Slim Jim vertical angle of radiation is almost parallel to ground so maximum radiation is where it is needed, straight out and all round. With all ground planes, including those with radials even one wavelength long, the vertical angle radiation is tilted upward at an angle of 30° or more. The shaded pattern in Fig. 2.3 is that from a $^5/_8\,\lambda$ ground plane with six quarter-wave radials. Compare this with the solid line pattern of the vertical angle radiation from a Slim Jim. Now examine the horizontal (omnidirectional) patterns in Fig. 2.4 both at a vertical angle of 0° i.e. a plane parallel to ground. The inner pattern shows the loss of radiation, by comparison, from a $^5/_8\,\lambda$ ground plane, a loss of the order of 6 dB. This has been verified with full size 2 metre antennas as well as with UHF scale models on the author's antenna test equipment (see Chapter 5).

How the Slim Jim operates

Basically it is an end-fed folded dipole operated vertically with voltage and current distribution as shown in Fig. 2.5(a). The matching stub provides a low impedance (50 Ω) feed at the base and couples to the antenna section at high impedance at one end. As with all folded dipoles, the currents in each leg are in phase whereas in the matching stub they are in phase opposition, so little or no radiation occurs from this. Correctly matched, the VSWR will be much less than 1.5 to 1 and will remain so across the band (see Fig. 5.2). The antenna can be constructed in a variety of ways for use at a fixed station or for portable operation.

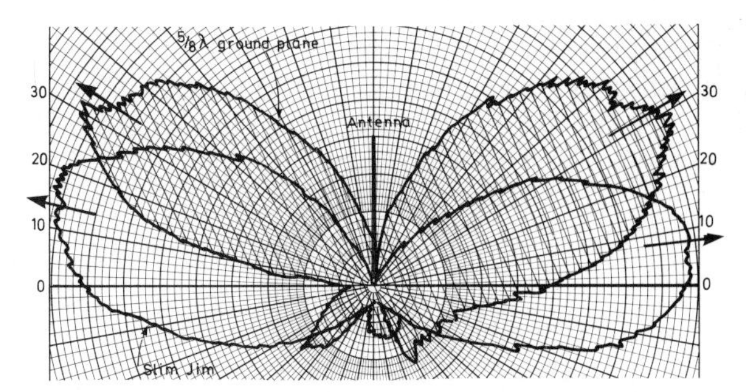

Fig. 2.3. Radiation at vertical angles from Slim Jim (solid line) and a $^5/_8\lambda$ ground plane (shaded pattern). From scale models on antenna range at 650 MHz

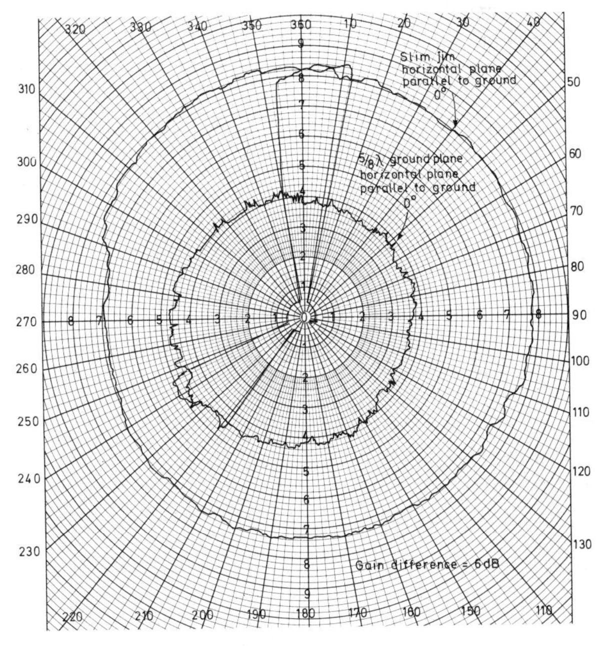

Fig. 2.4. Horizontal radiation patterns of Slim Jim compared with $^5/_8\lambda$ ground plane at 0°; parallel to ground. Gain difference at this angle is 6 dB. Patterns from 650 MHz models on antenna range

It has been used for mobile working mounted on a short stub mast attached to the rear bumper and, as on G2BCX/MM cruiser Frith 2, a special version is used completely enclosed in plastics tube for protection against salt water.

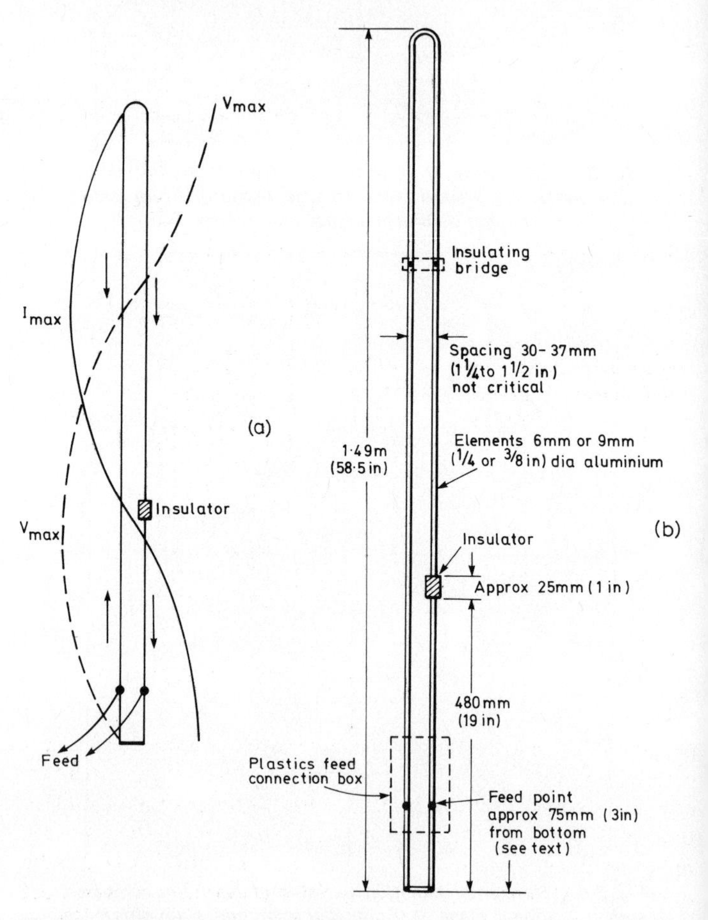

Fig. 2.5. Construction of the Slim Jim and its current and voltage distribution

Construction

The Slim Jim may be constructed from 6 mm or 8 mm ($^1/_4$ or $^3/_8$ in) diameter aluminium tube, stiff galvanised (coathanger) wire or 300 Ω ribbon feeder. The spacing between the parallel elements

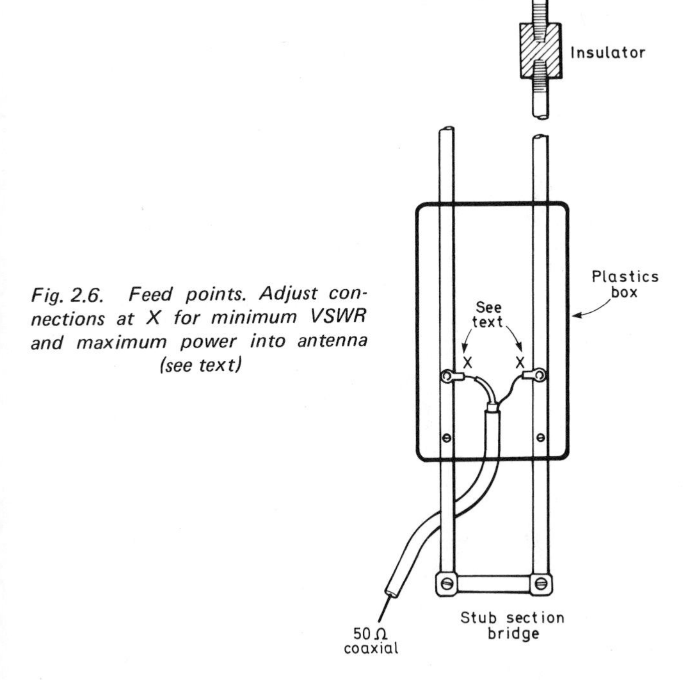

Fig. 2.6. Feed points. Adjust connections at X for minimum VSWR and maximum power into antenna (see text)

is not critical and neither is the overall length, provided this is within 6 mm (± $^1/_4$ in). The feed connection can be protected against weather by a small plastics box – an electrical junction box or similar – to which can be bolted a short wooden support for attachment to a mast. Insulation must be fitted between the return half of the folded radiator and the top of one side of the matching stub. This may be a piece of perspex or PTFE drilled to take the elements (they must not touch) which can be set in with a strong adhesive. If the aluminium tube has thick walls it may be possible to drill and tap them and use a ready made circular stand-off insulator with screw fittings.

The feed point may be protected from rain (as shown in Fig. 2.6) by a plastics box with a detachable lid. However, the correct feed point must first be found. Complete the construction of the antenna and stand upright near the transmitter but clear of

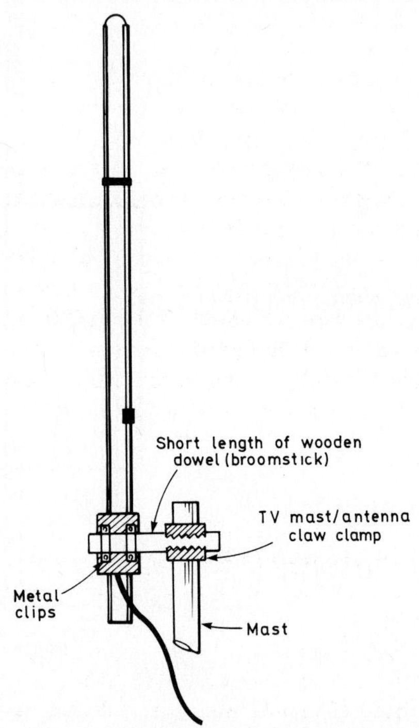

Fig. 2.7. Method of mounting the Slim Jim antenna on a mast

other conductors. Use the full length of feeder required to reach the antenna when finally *in situ*. Attach crocodile clips to the end of the coaxial cable, one to the outer braid and one to the inner conductor. Clip on at about 75 mm (3 in) up from the bottom as in Fig. 2.6. Adjust up or down for minimum VSWR and maximum power into the antenna. Note points of feed and fit solder tags as shown ready for soldering the feeder connections. The plastics box may now be fitted and it is suggested that the completed antenna and the feed protection box be given a coat or two of varnish. Fig. 2.7 shows a method of mounting on a mast with a TV antenna claw clamp. The antenna may be strengthened with a bridge of insulating material, perspex etc., across the elements as in Fig. 2.5(b).

Ideally the antenna should be as high as possible and clear of other conductors. It will however operate quite well indoors in a loft or even a living room but with shorter working range.

The 2BCX Slim Jim can also be constructed from galvanised iron wire or 300 Ω ribbon feeder. The long dimensions remain the same and the feed point adjustment is the same. The space between the elements may be reduced to about 25 mm (1 in). The whole antenna, made like this, could be housed in a length of plastics water pipe. If made from 300 Ω feeder the length remains the same and the insulator is not necessary. Simply cut about 25 mm (1 in) from the wire at that point and find the optimum feed point by removing insulation from the wires.

Ground plane antennas

Any antenna physically one quarter-wave long with reference to the frequency of operation is considered to be a resonant antenna and has a relatively low feed impedance if the feed point is at the base, i.e. at the end of the antenna when this is close to ground. Such an antenna radiates equally well in all directions around it and is therefore omnidirectional, and, assuming perfectly conducting ground as in Fig. 2.8, the vertical angle radiation is quite low in that it travels outward almost parallel to ground. However at VHF such an antenna would be at considerable disadvantage because at ground level and even with perfectly conducting earth beneath it, the radiation would be quickly absorbed by surrounding structures – buildings and trees etc. In any case real earth is far from being a perfect conductor so considerable loss in radiated power would occur because of this. The only answer therefore is to provide an 'artificial' earth of high conductivity material beneath the antenna to reduce the otherwise high ground loss.

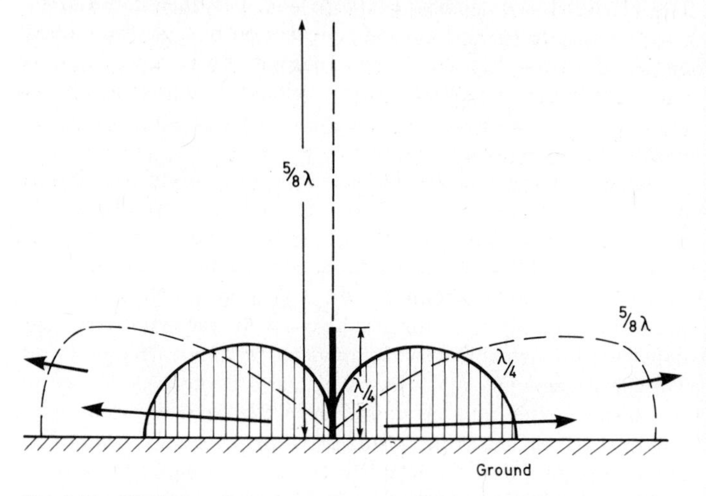

Fig. 2.8. Radiation from a vertical quarter-wave antenna at ground level by comparison with that from a $^5/_8\lambda$ vertical which, in theory but not necessarily in practice, yields a gain of about 3 dB

This still permits the antenna to radiate at relatively low vertical angles, and when the whole system is elevated as high as possible, attenuation by surrounding structures is largely overcome and the point of horizon and therefore working distance is increased.

A $^5/_8\lambda$ ground plane antenna

One of the most popular omnidirectional antennas, although not necessarily the most efficient, is the so-called $^5/_8\lambda$ ground plane antenna, and for fixed station operation this usually consists of the antenna itself mounted above a system of metal radials which form an 'artificial' earth or ground plane as previously described. This antenna is also used for mobile operation in which case it is mounted on the roof, or some part of the body of the car, which acts as the ground plane.

It is called the $^5/_8\lambda$ ground plane because this is approximately its physical length. Electrical length is three-quarters of a wavelength but as the current and voltage distribution along the antenna shows (see Fig. 2.9), the first section is electrically a quarter wavelength (but physically about one eighth because of the loading coil). The feed point impedance is about 50 Ω. Construction is not difficult, and both the antenna and its radial section can be made from thin copper or aluminium tubing, from stiff wire

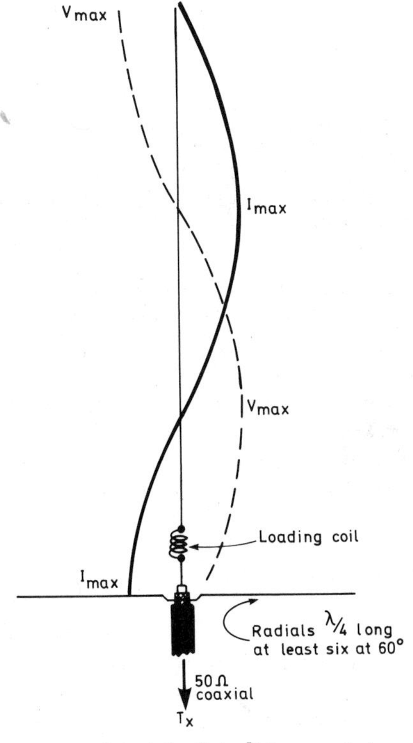

Fig. 2.9. Principle of the $^5/_8\lambda$ ground-plane antenna

such as welding rod, or even from wire coathangers straightened out. The ribbed frame from an old umbrella also makes a very efficient (and if required a folding) ground plane, as will be shown later. The loading coil consists of four turns of 16 swg (1.63 mm

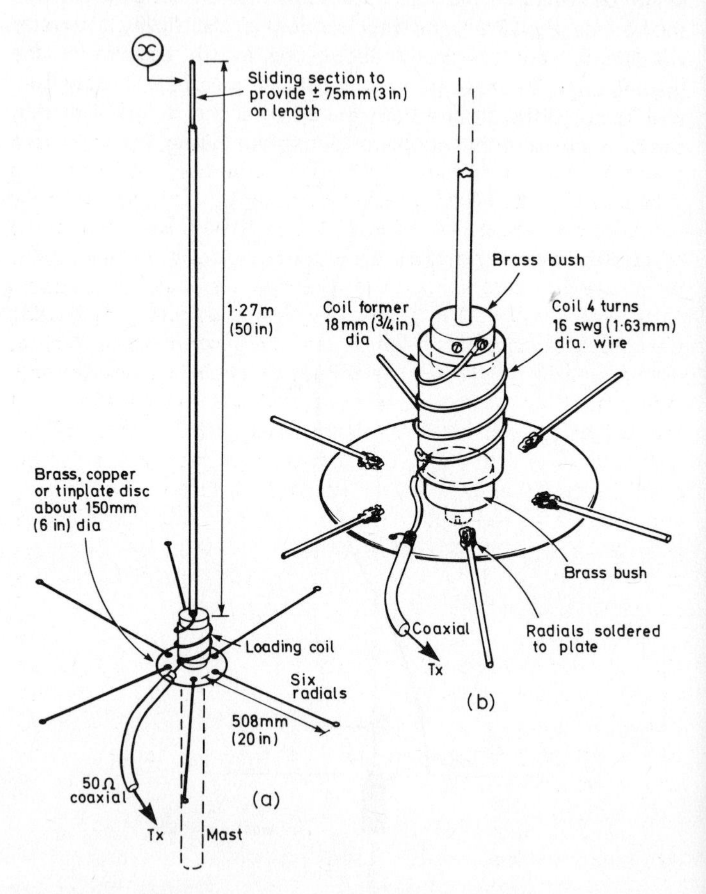

Fig. 2.10. (a) Dimensions for the construction of a $^5/_8\lambda$ ground-plane antenna. (b) details of loading coil and radial assembly for the $^5/_8\lambda$ ground-plane antenna

diameter) on an 18 mm (3/4 in) diameter insulating former of paxolin or PTFE, and the top portion of the antenna is made adjustable to achieve resonance and low VSWR. An outline for construction is given in Fig. 2.10.

There should be no need to trim the length of the 50 Ω coaxial feed cable as is often wrongly done, and with the radiator properly adjusted the VSWR should be at least 1.1 or 1.2. With 10 W going into the antenna it should be possible to fully light a small 6 W, 240 V fluorescent tube at the voltage points marked (X).

A proposed method of constructing the loading coil and radial assembly is given in Fig. 2.10(b). The radials should be not less than 508 mm (20 in) long but may be longer, and, whilst four are essential, six would be better. The base mounting plate may be of copper, brass or tinplate thick enough to support the radials which are bolted or soldered to it. Do *not* bend the radials down to achieve a low VSWR. If this cannot be obtained by adjusting the length of the radiator there may be another reason such as wrong feeder impedance or a mismatch at the transmitter end. Adjustment for resonance can be carried out with the antenna at low height but clear of other conductors, and if it is to be used outdoors make sure that water cannot enter the open end of the coaxial cable where it joins the coil and where the outer braiding ends. Seal with Sealastic or similar non-conducting sealing compound. The coil can be covered with PVC tape and the whole antenna and the coil given a coat or two of polyurethane or similar varnish for protection against weather.

This antenna is also most suitable for mobile operation using the car body as a ground plane. Centre rooftop mounting is best, but the loading coil must be insulated from the car metalwork. The 50 Ω feed cable is the same except that the screening braid is connected to the car metalwork as close to the base of the antenna as possible.

The umbrella ground plane

As mentioned earlier an old umbrella frame, shorn of its cover, makes an excellent ground plane for 2 metre operation. The ribs will need to be connected together using old coaxial cable outer braid, as electrical contact between them is rather poor. Using the $^5/_8\,\lambda$ radiator and loading coil as in Fig. 2.11, the umbrella ground plane makes a useful portable entenna, particularly if the radiator is made detachable. The ground plane can be folded up. For portable use all that is required is a light sectional mast. An arrangement used by the author is shown in Fig. 2.11, and with

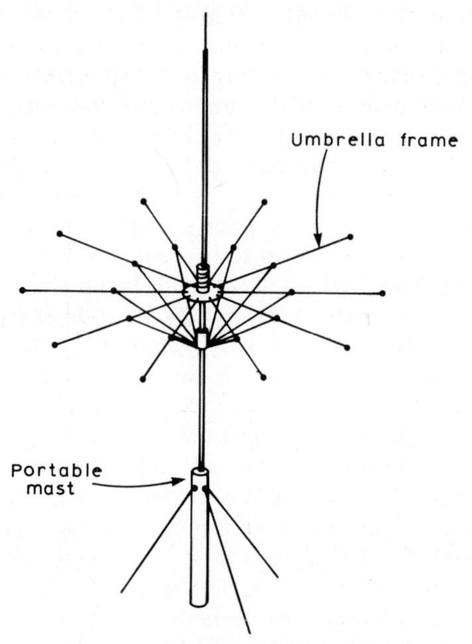

Fig. 2.11. An old umbrella makes an excellent ground plane for portable work when used in conjunction with the radiator described in Fig. 2.10 and a small sectional mast

this in clear surroundings but at height of only 4 m (12 ft) quite considerable DX was worked during tropospheric lift conditions. *NOTE*. The $^5/_8\,\lambda$ ground plane has a gain of approximately 3 dB over a dipole but *only* at the optimum angle of radiation in the vertical plane as referred to previously in Fig. 2.3 and associated text.

Collinear antennas

These are among the very few antennas that in the vertical mode are omnidirectional, that can be made to provide a useful degree of gain over a dipole and that are not too difficult to construct. There are various commercially made collinears available for 2 metre operation, but the claims made for gain should be viewed with some suspicion. For example, the specification may quote '10 dB gain' (which is meaningless anyway unless a reference is given) but this may be relative to the isotropic radiator, a hypothetical aerial which is regarded as having no gain and used as a

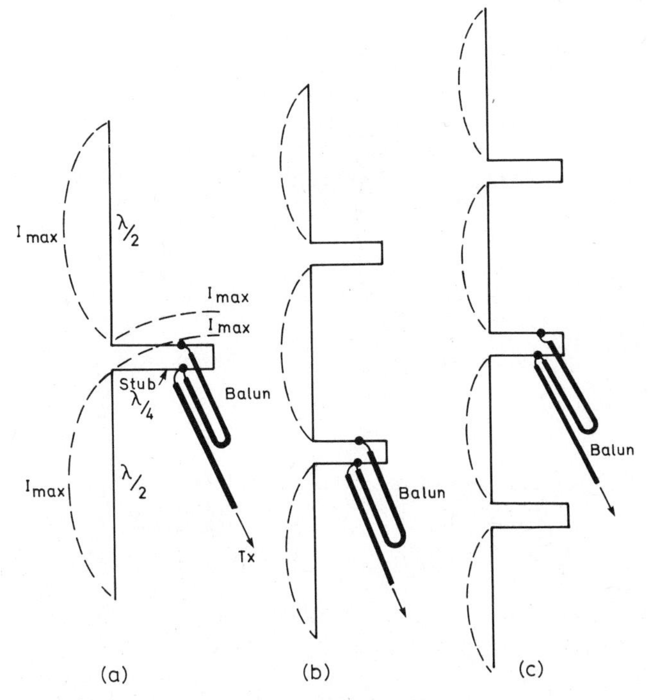

Fig. 2.12. (a) Two-element collinear. (b) Three-element collinear. (c) Four-element collinear. Current distributions as shown by broken lines

0 dB reference. The standard 0 dB reference aerial for gain is the dipole in free space (which has a gain of 1.64 dB over an isotropic), but first let's see what real gain can be obtained from collinear systems and also how they operate.

A simple collinear array consists of two half-wave radiators made to operate 'in phase' so the total radiated power is about 1.5 times that from a single dipole, i.e. the power gain is about 2 dB. The physical configuration of a two-element collinear together with its voltage and current distribution is shown in Fig. 2.12(a). More elements can be used to achieve higher gain and with the conventional close spaced element system this will be about 3.2 dB from a three-element system and about 4.3 dB from a four-element system. More than four elements are rarely used as the increase in gain is not worth while. Gain can be increased by spacing the elements wider apart, and with a spacing of about 0.4 λ a gain of more than 3 dB can be obtained with

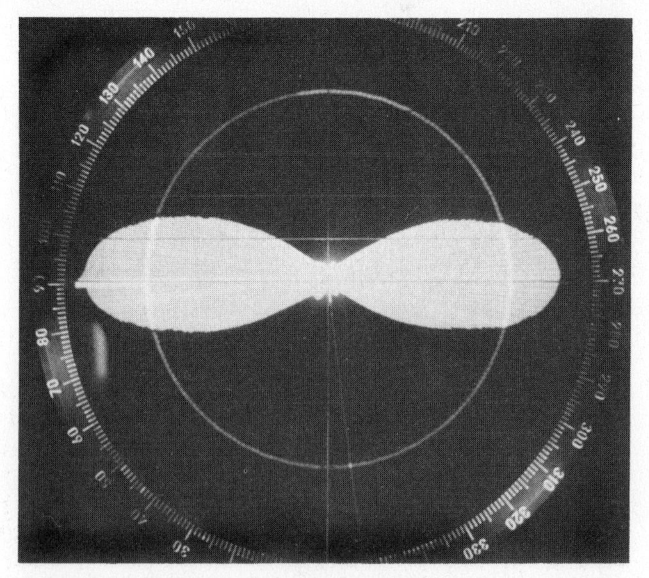

Fig. 2.13. Vertical angle radiation from a 2-element collinear. Horizontal mode radiation is similar

only two elements, but then it becomes very difficult to achieve a suitable feed match and phasing with 50 Ω (or 75 Ω) cable. The closely spaced element system is better from the constructional and feed point of view. The vertical angle of radiation from a two-element collinear is shown in Fig. 2.13, but like all VHF vertical aerials it needs to be high up, the higher the better.

Construction

First the whole antenna must be well insulated from the mast or other support, the reason being that the ends of the elements are at high r.f. potential, which means that insulation between elements must be good as well. The elements and matching stubs can be made from large diameter copper wire but this would require extra insulating support. Copper, brass or aluminium tubing is better and also more or less self supporting. Tubing made from PTFE or similar low loss material can be used for linking lightweight elements, or one could of course do a complete 'plumbing' job with 12 or 15 mm ($^1/_2$ or $^5/_8$ in) diameter copper water pipe for the elements and stubs and using appropriate fittings to join them together. The author has built quite a number of collinears using all the materials mentioned, but successful construction often depends largely on ingenious use of readily available materials.

Feeding and matching

Before setting about the construction however, we should examine the method of feeding a collinear antenna from 50 Ω (or 75 Ω) coaxial cable. Since the radiating elements must be in phase, they are connected together by means of quarter-wave stubs, the whole system being fed at a suitable point on one of the stubs. Coaxial cable however, provides an unbalanced feed, but by using a balun, as shown in Fig. 2.14, an impedance step up of about 4:1 can be obtained, i.e. to about 200 Ω, as well as a balanced connection. Provided the elements are resonant and the stubs properly constructed, good matching and low VSWR can be obtained by finding the right tapping points along the stub for the feeder/balun connection. The configurations for 2-, 3- and 4-element collinears are shown in Fig. 2.12.

A two-element collinear

This is sometimes referred to as the 'plumbers delight' as the elements and stub are made from copper water pipe joined with appropriate fittings. Note that a blow lamp is needed to soft solder the fittings.

The antenna is self supporting and can be mounted on the top of a metal mast as shown by means of a T junction let into the lower leg of the stub and coupled to the mast by a length of PVC waste pipe. Suitable couplers can be obtained for this from builder's merchants. All the dimensions are given in Fig. 2.14(a) and details for the balun feed will be found in Fig. 2.14(b). Adjust the tapping points (X) for maximum radiation with the aid of a field strength meter, series power meter, neon or fluorescent tube at voltage points etc., and also for low VSWR which should not be greater than about 1 to 1.2. Adjustment can be made on a received signal, i.e. adjust for maximum signal meter reading.

The balun feed loop is effectively a half-wavelength long but because of the velocity factor of the coaxial cable (same as used for the feed cable itself) it will be physically shorter, average 712 mm (28 in). The following simple formula will give the exact length for the type of coaxial cable used (e.g. UR43, UR67 etc.):

$$\text{length (m)} = \frac{150 \times \text{velocity factor of cable}}{145\ \text{MHz}}$$

The tapping points along the stub will be around 150 mm (6 in) from the closed end.

A Slim Jim collinear

This has been tested as a scale model at 650 MHz and as a full size version at both 145 and 430 MHz and is an extension of the Slim Jim described earlier. With any single-element close spaced collinear

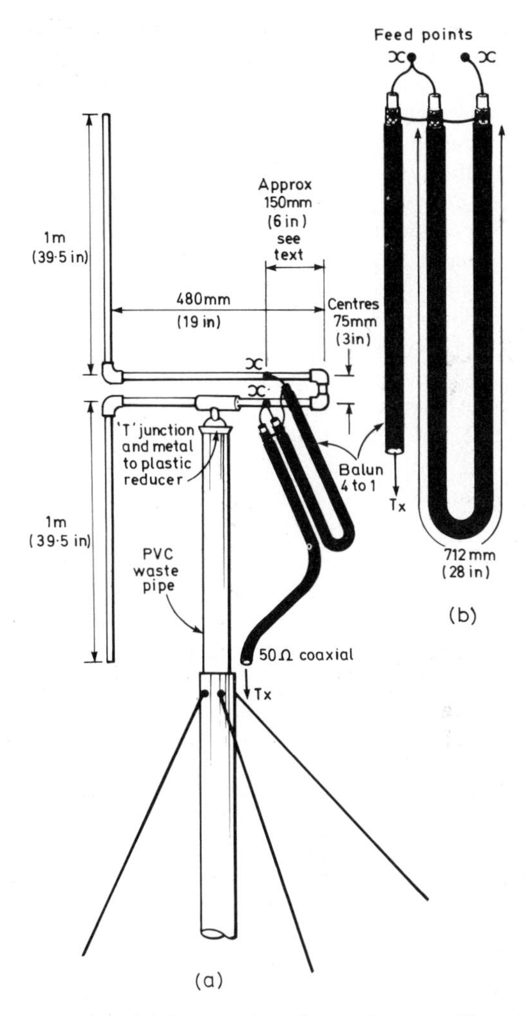

Fig. 2.14. (a) Construction of a two-element collinear from copper water pipe and fittings. (b) Details for the balun (see text)

system the gain over a dipole is about 2 dB, but the use of folded elements permits a gain of about 3 dB. The principle of operation is the same as the single-element Slim Jim but in addition there is an extra radiator coupled by means of two $\frac{1}{4}\lambda$ phasing stubs so that both elements operate in phase as in Fig. 2.15. For construction refer to Figs. 2.5 and 2.6 and from this it is only a case of making the system self-supporting perhaps with 9 mm ($\frac{3}{8}$ in) or smaller diameter tube, or say 16 swg (1.63 mm diameter) copper wire, arranged on a light wood frame on stand-off insulators. For resonance at 145 MHz it may be necessary to make the upper element slightly shorter than $\frac{1}{2}\lambda \times 0.95$ which is the length of the lower element. The base or driving stub is 480 mm (19 in) long but the two phasing stub lengths may have to be found experimentally by using adjustable shorting bars at each end and making adjustment to obtain the lowest possible VSWR, i.e. less than 1.5 to 1. The optimum length of these stubs will be (a) about 468 mm ($18\frac{1}{2}$ in) and (b) 456 mm (18 in), with the top folded element about 960 mm (38 in) long to which adjustment could also be made with a movable shorting bar at the top to obtain exact resonance. When the stub lengths have been settled they could be re-made into loops as shown by the broken lines in Fig. 2.15, if only to make the antenna look a little less religious.

Mobile and portable operation

There are many commercial mobile antennas for 2 metre operation, and most are $\frac{1}{4}\lambda$ or $\frac{5}{8}\lambda$ types for side or roof mounting. Probably the most efficient is the $\frac{5}{8}\lambda$ type when mounted centrally on the roof top so that the roof itself, providing it is all metal, offers the largest possible ground-plane area. Antennas mounted at sides front or rear, do not behave very efficiently because of the irregular ground-plane area, close proximity to other parts of the car body and ground itself. It is also very important if the aerial is to provide maximum radiated power, to have the outer (screen) of the cable bonded at the point where the antenna is fitted. Magnetic mounts for example may not always provide a low enough

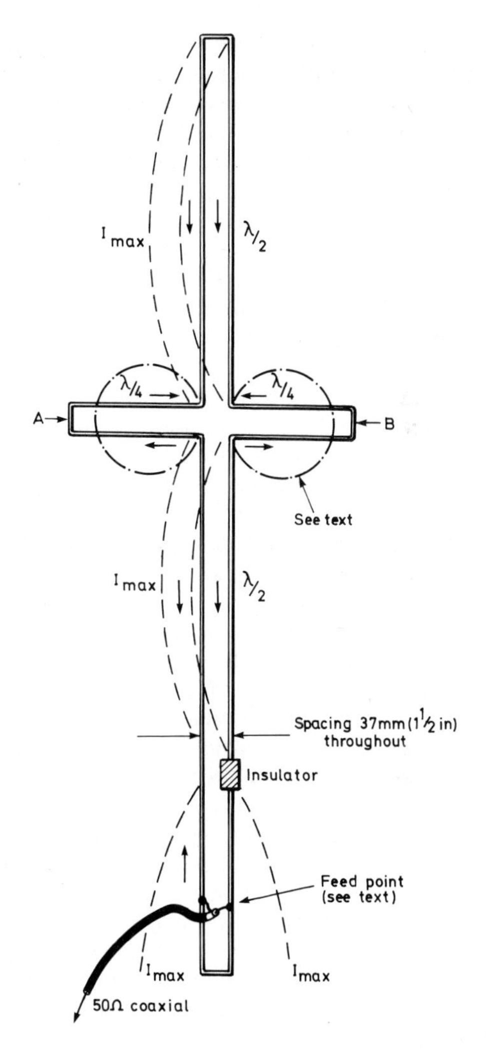

Fig. 2.15. The Slim Jim collinear, suitable for 2 metre or 70 cm operation (see text for construction)

resistance 'earth' connection for the cable or indeed any earthing connection at all, the only earth the cable 'sees' being the capacitance between the magnetic mount and the car body with paint work as a dielectric. Antennas $^1/_4\lambda$ long mounted at sides, i.e. wing mounting, are not only inefficient in terms of ERP but are even more shielded by the car bodywork than would be the much longer $^5/_8\lambda$ where the upper current carrying portion is more likely to be clear of bodywork. Adapted car radio aerials are not very efficient either for much the same reason.

There are no 'special' antennas for mobile operation that will provide both gain and omnidirectional coverage, except the $^5/_8\lambda$ which is supposed to give 3 dB more radiation than a dipole but really doesn't in the vertical angles near ground, i.e. between 0 and about 30°. With the $^1/_4\lambda$ type most of the power goes straight up unless it is mounted at roof centre. A bumper-mounted two-element collinear might be thought a possibility for more efficient operation if one doesn't mind being conspicuous, but tests carried out have proved this *not* to be the case as the antenna is physically too close to real earth. There is, however, a form of collinear that has been successfully used by the author and this is a $^3/_4\lambda$ collinear in which the first section of $^1/_4\lambda$ is operated in the same phase as the $^1/_2\lambda$ section above it (Fig. 2.16(a). With a $^5/_8\lambda$ antenna the lower $^1/_4\lambda$ portion of the element, i.e. the loading coil plus about $^1/_8\lambda$ of element, operates in phase opposition to the main $^1/_2\lambda$ section so a small amount of radiation is lost at very high vertical angles. Bringing both sections of the antenna into phase ensures that all the radiation goes in one direction which theoretically is virtually parallel to ground. However, the ground-plane area provided by the metal top of a large car is much the same as for other vertical antennas in that it is not large enough to keep the radiation at a very low angle. For those who might like to try the $^3/_4\lambda$ collinear as an alternative to the $^5/_8\lambda$ antenna, details are given in Fig. 2.16. The $^1/_4\lambda$ phasing stub (c) can be coiled up (d) and (e) so the antenna will remain fairly unobtrusive. It can be fed directly from 50 Ω coaxial cable but some adjustment may be required to the length of the top section and the rolled up stub to achieve a low VSWR. This antenna can also be used with a conventional ground plane.

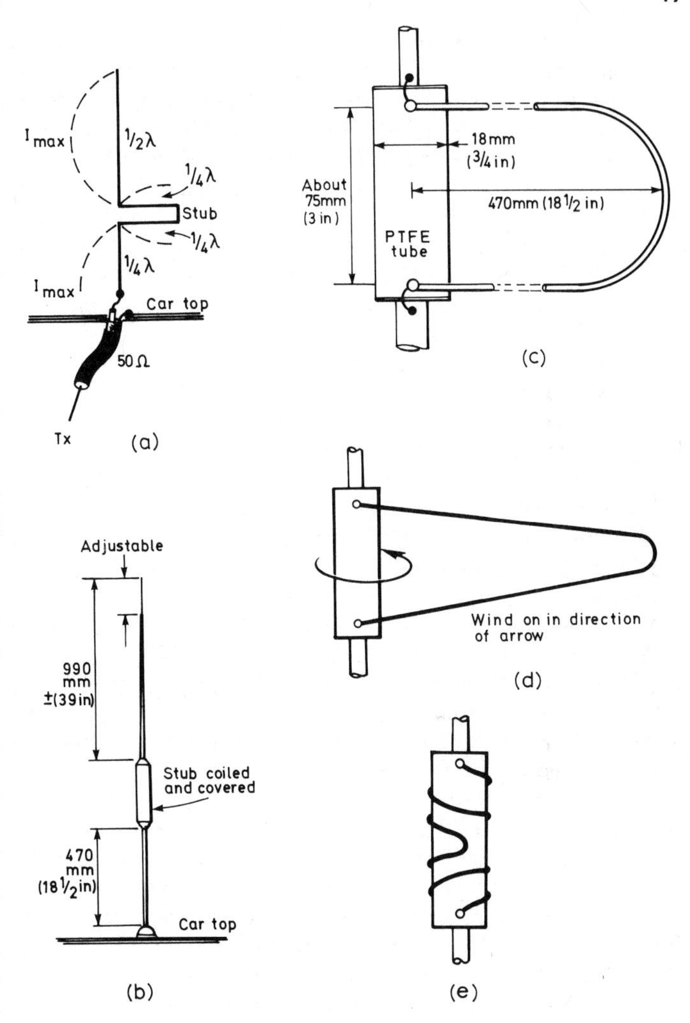

Fig. 2.16. Details for constructing a three-quarter collinear for mobile operation. (a) current distribution; (b) suggested construction; (c) first assembly of phasing stud; (d) before winding the stub on the former; (e) the stub wound onto the former

Antennas for hand-held tranceivers

The antennas either supplied with, or sometimes built in to, hand-held portable tranceivers are probably the most efficient that can be used for the purpose intended. There is little to choose between the $^1/_4\,\lambda$ pull-out rod and the small helically wound types sometimes used. Maximum radiation from such antennas is upwards, i.e. off the end of the antenna, and could only be directed more toward the horizon by using a large ground-plane which would of course defeat the object of the compact hand-held transceiver.

Antennas for portable operation

Virtually any antenna can be used for fixed site portable operation, for example, antennas used on holiday or at field days etc. Making the antenna reasonably small and easy to transport are the primary requirements. The construction of an otherwise physically large antenna is generally a matter of employing one's ingenuity to make it fold up or be reduced to sections that can be put together quickly. Small entennas like the ZL Special, two-element version, and the HB9CV will fit straight into the boot of a car. High gain multi-element beams could be made up with joints in the boom so that the reflector and driven elements are on one section for example, and with two or three directors per section of boom for the remainder.

The umbrella ground plane described in Fig. 2.11 also makes a useful portable antenna, and instead of a $^5/_8\,\lambda$ radiating element the $^3/_4\,\lambda$ collinear described in Fig. 2.16 could be used instead. Another omnidirectional that could be sectionalised is the Slim Jim or its collinear version (Figs. 2.5 and 2.15).

A lightweight portable mast could be made from, say, 1 metre (3 ft) lengths of 25 mm (1 in) diameter dowel (broomsticks) joined together with metal sleeves and supported from three or four thin nylon guy lines. Four or five sections would be about the limit for a lightweight mast of this type.

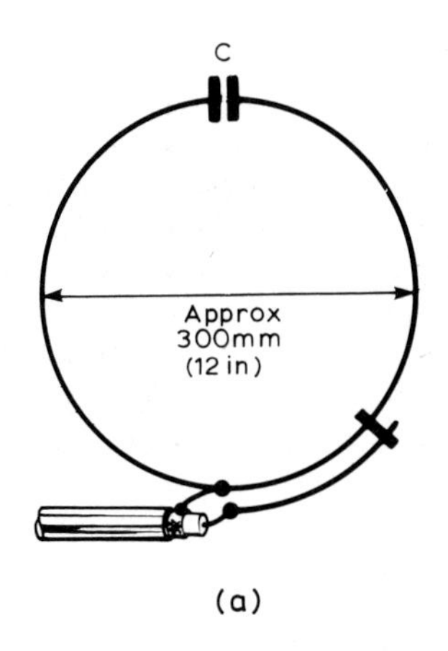

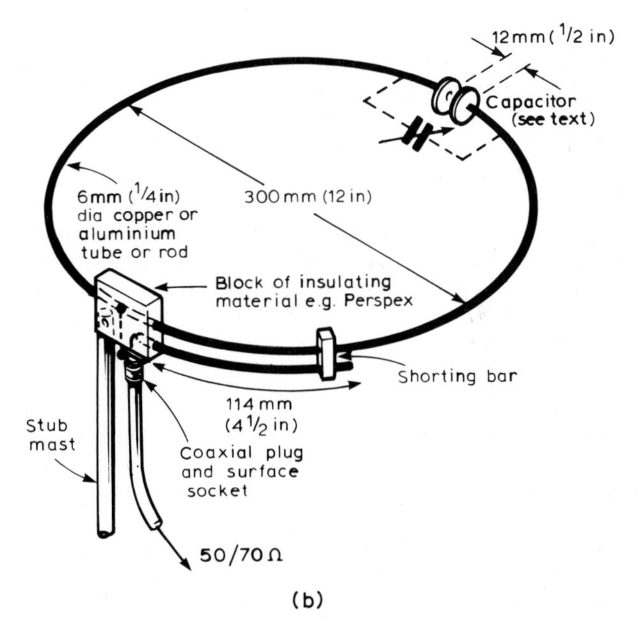

Fig. 2.17. Details for construction of a halo antenna – see text for details of adjustment

Other omnidirectional antennas

For omnidirectional operation with *horizontal* polarisation the 'halo' is probably the best known, and details of its construction are given in Fig. 2.17. It is worth considering for mobile SSB working as many SSB operators wisely use antennas that are horizontally polarised and by doing so vastly increase working ranges.

The sleeve dipole is another simple vertical antenna but it is not particularly efficient. An improvement on this is a coaxial stub fed full half-wave, although the J match half-wave or the Slim Jim will function just as efficiently. A collinear constructed from sections of coaxial cable is described in the *ARRL Antenna Handbook*, but is rather difficult to make and, moreover, to operate efficiently. It also requires external support by means of a wood or plastics mast, or by placing it within a plastics tube.

For omnidirectional radiation with horizontal polarisation there are also the crossed dipole systems, the turnstile antenna, the horizontal loop, which is similar to the halo already mentioned, the clover leaf, which consists of four small horizontal loop antennas fed from a single coaxial cable, and the square or 'Alford' type consisting of four $^1/_4\lambda$ elements fed so that the phase of current in each is the same. These antennas are not too difficult to make and information concerning design can be found in the reading list given at the end of this chapter.

The halo antenna

The halo was at one time commonly used for mobile operation when most transmitting and receiving was done with horizontally polarised antennas. It would still be a useful omnidirectional antenna for SSB operation. The halo is in effect a dipole formed into a ring and gamma matched from 50 or 70 Ω coaxial cable. As Fig. 2.17 shows, the diameter of the ring is about 300 mm (12 in) and the length of the gamma section about 114 mm ($4^1/_2$ in). Matching is obtained by moving the gamma shorting bar while tuning by means of the capacitor, both being adjusted for minimum VSWR. For initial adjustment an air spaced variable capacitor of about 50 pF could be used. When minimum VSWR has been obtained, the capacitor can be replaced by the discs, made of thin aluminium spaced about 12 mm ($^1/_2$ in) apart and these are trimmed to produce the same VSWR.

Stacked loop antenna

This is also horizontally polarised and omnidirectional and being considerably more efficient than a halo is more suitable for fixed site operation. The distance around each loop is 1λ, so each loop contains 2 × $^1/_2\lambda$ radiating sections. One loop alone has a field pattern the same as that of a dipole, but by using two stacked loops at right angles to each other phased 90° by the $^1/_4\lambda$ connecting line, an almost perfect omnidirectional pattern is obtained, as shown by the polar pattern in Fig. 2.18 which is from a 650 MHz

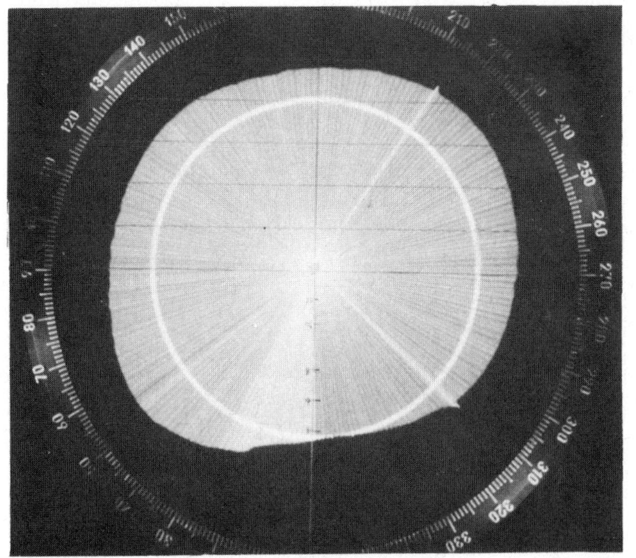

Fig. 2.18. Omnidirectional radiation pattern from a 650 MHz model of the horizontally stacked polarised loop antenna, as in Fig. 2.19

model. Each loop is identical in size and $1/2\lambda$ can be taken as approximately 990 mm (39 in). The arrangement is shown in Fig. 2.19. The feed point should be made via a matching stub or balun with a 4 to 1 ratio if 50 Ω coaxial cable is used as the main feed. A balanced feed is essential. The gain of this antenna should be in the region of 3 dB over a single dipole.

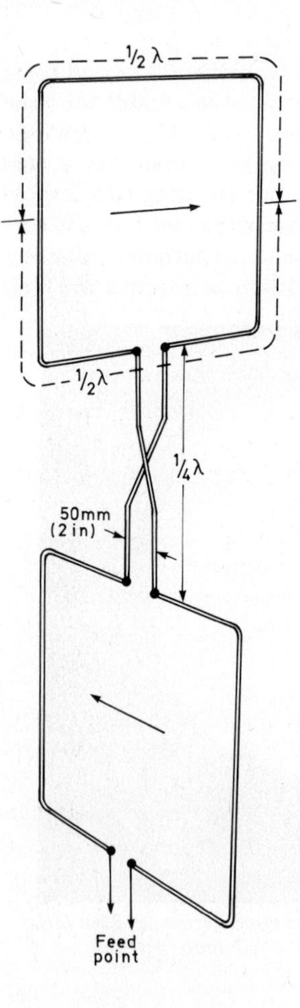

Fig. 2.19. Design for stacked and phased pair of 1λ square loops. Radiation is omni-directional and horizontally polarised

Omnidirectional folded element collinear

A conventional collinear antenna with two in-phase close spaced elements has a gain of about 1.8 dBd. Higher gain from collinear pairs can be obtained only by wider spacing between the elements, but this entails a rather complex feed system to maintain the necessary phasing. The highest gain that can be obtained by this method is about 3 dBd with the elements spaced 0.4 λ apart.

The antenna described here was designed by the author for a 70 cm repeater station and employs two close spaced half-wave elements, but each is a folded dipole and the resultant gain is a little over 3 dBd. Phasing is achieved in the usual way with a quarter-wave stub, although the diagram shows a stub of 0.625 λ in length which is to obviate, to some extent, the problem of distorting the otherwise omnidirectional field pattern when the antenna is operated in vertical polarisation mode and is attached to a metal mast.

A mounting distance of 0.625 λ from a metal mast has been found about optimum in keeping reflection from the mast at minimum, which would otherwise reduce radiation in the direction behind the mast and antenna. *This critical mounting distance does not guarantee freedom from this problem, and if a true omnidirectional field is required it is recommended that a mast of insulating material (wood or PVC pipe) is used, extending 300 mm (1 ft) or so above and below the antenna.* The vertical angle of maximum radiation is parallel to ground, i.e. at zero degrees, and the lobe width at –3 dB (half power point) is in the region of 60°. The antenna has a wide bandwidth and should maintain a VSWR to around 1.1 or 1.2 to 1 across either the 70 cm or 2 metre band. Low loss semi-airspaced 50 Ω coaxial cable should be used for 70 cm working, but unless the cable run is very long, lower grade cable such as UR43 could be used for 2 metres. Ideally the lowest loss cable that can be afforded should be used with any aerial. Remember X dB power loss in the feeder is also X dB loss in radiation.

The elements and stub may be constructed from 6 mm ($^1/_4$ in) diameter aluminium rod which is easy to drill and tap M3 (4BA) at the ends of the elements to accommodate the bridging pieces.

The format is shown in Fig. 2.20 and dimensions for the element and phasing stub lengths are as follows:

	Sections	*70 cm*	*2 metres*
(a)	Half-wave elements	324 ± 3 mm ($12^3/_4$ ± $^1/_8$ in)	960 ± 6 mm ($37^3/_4$ ± $^1/_4$ in)
(b)	Total stub length	390 ± 3 mm ($15^3/_8$ ± $^1/_8$ in)	1168 ± 6 mm (46 in ± $^1/_4$ in)
(c)	Elements and stub section spacing (centre to centre)	19 ± 3 mm ($^3/_4$ ± $^1/_8$ in)	25 ± 3 mm (1 in ± $^1/_8$ in)

The above dimensions take into account velocity and capacitive factors

Note the insulating block between the ends of the folded elements where they meet opposite the feed point. Material must be of good quality such as PTFE or equivalent.

The mast fixing bracket at the end of the stub may be 12 mm ($^1/_2$ in) wide by 1.5 mm ($^1/_{16}$ in) or 3 mm ($^1/_8$ in) aluminium about 150 mm (6 in) long and drilled either side of the stub termination for 6 mm ($^1/_4$ in) diameter fixing bolts. The quarter-wave point shorting bar can be made from aluminium about 9 mm ($^3/_8$ in) square and drilled to take the stub lines and also drilled and tapped M3 (4BA) at the ends to lock to the lines when adjustment is completed.

When the antenna is assembled it should be set up in fairly clear surroundings about 2 m (6 ft) above ground with the full length of feed cable to be used already attached. Adjust the feed tapping points and quarter-wave point shorting bar together for maximum power and/or minimum VSWR. The feed point will of course require protection from rainwater and to this end a small oblong plastics box could be used to enclose the feed connections and the end of the cable. It is important to have a good seal to prevent water getting in at this point. If water seeps into coaxial cable it will ruin it for all time.

All VHF aerials operate most efficiently when high up above rooftops and other conducting obstacles such as tall trees. This applies especially to aerials such as vertical collinears with low (zero angle) radiation. Sizeable trees in full leaf in the path of radiation can attenuate radio waves at VHF and UHF by as much as 20 dB, even in dry weather.

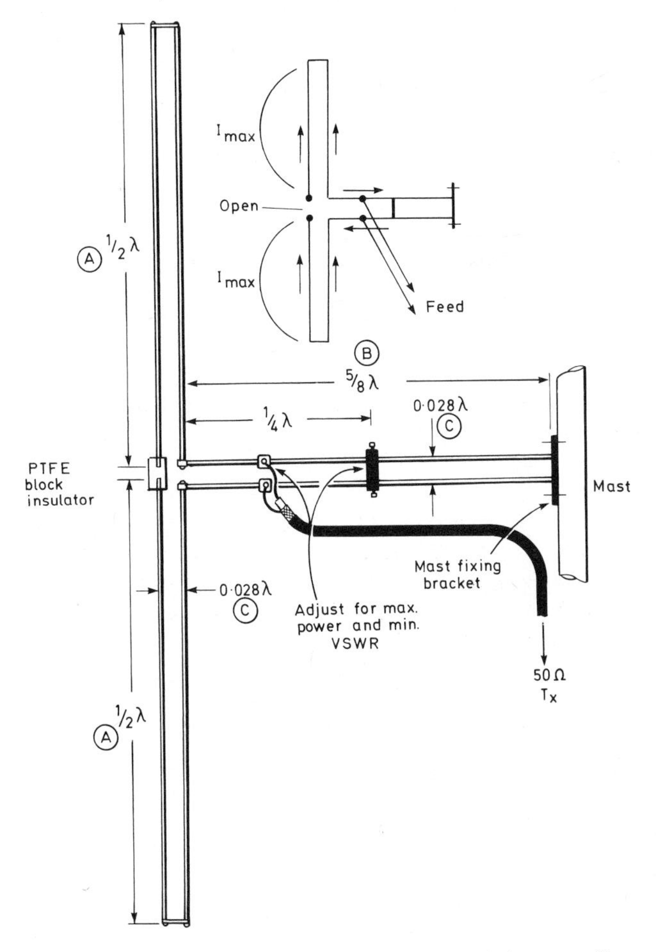

Fig. 2.20. Details for the construction of a folded element collinear antenna for 2 m or 70 cm operation

Coaxial stub fed vertical

This antenna is a modified Slim Jim but it was originally developed for marine use and is therefore totally enclosed for protection from weather and salt water. It would be a fairly inconspicuous antenna for fixed station or mobile use (with stub mast and bumper mounting) or a 2 metre repeater station antenna. Although very efficient it is more difficult to construct than the original Slim Jim because of the ¼λ long copper tube coaxial feed and very closely spaced elements. The radiating portion is a folded

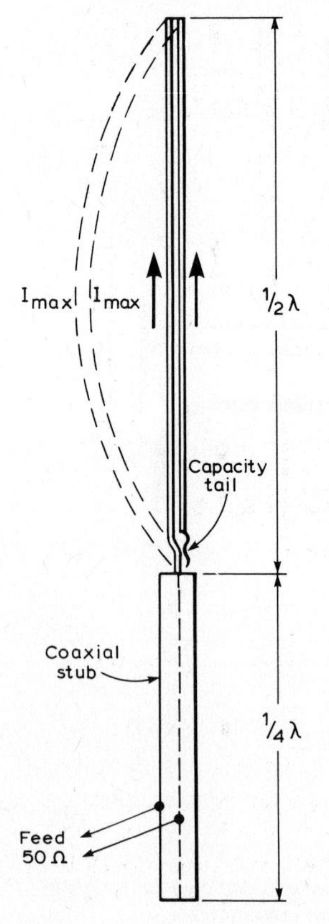

Fig. 2.21. Arrangement for end feeding and a folded half-wave element with a quarter-wave coaxial stub

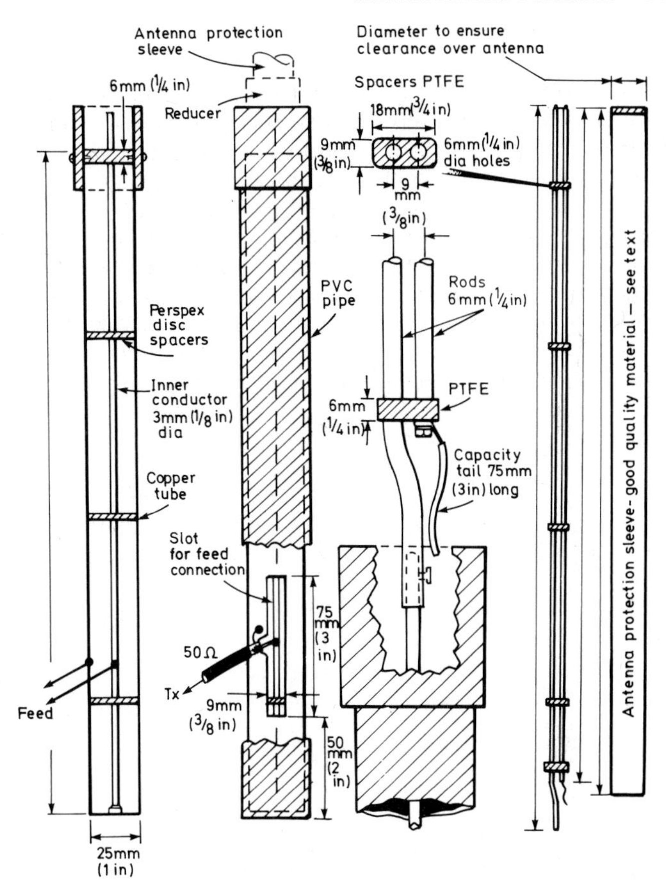

Fig. 2.22. Constructional details for a fully enclosed coaxially fed folded half-wave antenna, as in Fig. 2.21

dipole, end fed from the coaxial stub which itself is directly connected to a 50 Ω feed cable. The general arrangement is shown in Fig. 2.21, and Fig. 2.22 includes all dimensions and details for the coaxial $\frac{1}{4}\lambda$ stub and radiating elements. The stub is made from 25 mm (1 in) diameter copper tube sealed at one end with a metal (copper or brass) disc soldered in which forms the short circuit and connection for the inner conductor. At least four or five circular Perspex spacers about 3 mm ($^1/_8$ in) thick are required to keep the inner conductor straight through the centre of the stub but the one at the top should be at least 6 mm ($^1/_4$ in) thick as shown. A slot about 9 mm ($^3/_8$ in) wide and 75 mm (3 in) long starting about 50 mm (2 in) up from the bottom of the stub must be cut to allow access for the feed cable and adjustment of the tapping point to ensure minimum VSWR. The radiating elements are made from 6 mm ($^1/_4$ in) diameter aluminium rods connected at the top end by a metal bridge and held apart by PTFE spacers (at least five) along the length as shown. Final adjustment for operation is made by first setting the feed point tapping to the coaxial stub inner conductor for the lowest possible VSWR. Further adjustment can be made with the capacity tail which may be folded back or altered in length until the VSWR comes down to about 1.1 or 1.2 to 1. Note that the capacity tail, which may be of 18 or 16 swg (1.22 or 1.63 mm dia.) wire, must have an insulating sleeve over it to prevent contact with any other part of the antenna. The sleeve should preferably be of PTFE. Radiation from this antenna is of course omni-directional and on a plane virtually parallel to ground.

[*PTFE* (Polytetrafluorethylene). A special low loss insulating material (teflon or fluon) suitable for use at VHF and UHF.]

Reading list

Subject	*Reference*
Omnidirectional antennas, VHF and UHF	*ARRL Antenna Book*, American Radio Relay League, available in UK from Radio Society of Great Britain, 35 Doughty Street, London, WC1N 2AE.
As above	*Radio Amateurs VHF Manual*, American Radio Relay League, available from R.S.G.B. as above.
As above	*VHF–UHF Manual*, G. R. Jessop, R.S.G.B. as above.

Mobile antenna for magnetic mounting	*Ham Radio Magazine*, (USA), September, 1975.
$^5/_8\,\lambda$ vertical antennas	*Ham Radio Magazine*, (USA), May, 1974
Antennas for VHF/UHF	*Ham Radio Magazine,* (USA), May, 1973.
Vertical antennas for for 144 MHz	*Practical Wireless*, (UK), July, 1976.
$^5/_8\,\lambda$ antennas for mobile operation	*Ham Radio Magazine*, (USA), May, 1976.
The Slim Jim omni-vertical	*Practical Wireless,* (UK), April, 1978, *CQ-PA*, VRZA, Dutch Amateur Radio Society, June, 1978.

Chapter 3

Directional Antennas

The previous chapter dealt with vertically polarised omnidirectional antennas, suitable for general coverage and working mobile and repeater stations, most of which employ vertically polarised antennas. With the exception of the collinear arrays these antennas have little or no gain over a dipole, so in order to generate a greater ERP (effective radiated power) one must resort to directional antennas.

Remember, there are directional antennas (and the dipole, when used in horizontal mode, is one of them) that do not provide an increase in radiated power in one or more directions – such antennas have no *directivity gain* unless compared with an isotropic radiator. Gain in radiated power from an antenna is therefore sometimes called directivity gain. There are of course many antennas to choose from, with the Yagi being perhaps the most popular, although there is much controversy over the question of overall length and director spacings relative to the obtainable gain. Sources of further information on this will be found in the reference list at the end of the chapter. In any case, to achieve a high gain with a Yagi array means an antenna either of considerable length, or a stacked and/or bayed system of two or more. This is not very practicable for sites with limited space, and some may even be prohibited from having an outside antenna of any description.

End-fire arrays

Fairly small but efficient beam antennas for 2 metres (or the UHF 70 cm band) are not difficult to make, and a directivity gain as high as 10 dB is possible without having to use long arrays like the Yagi, or similar, with large numbers of elements. The corner reflector may be an exception, but although a gain of around 12 dB is possible, it is of somewhat large dimensions.

Small beams with a useful degree of gain intended for operation in vertical or horizontal mode may be derived from what are usually called 'end fire' arrays and also from a similar arrangement known as a 'broadside' array, although the latter has less practical application as far as 2 metre operation is concerned. Such arrays consist basically of two spaced vertical (or horizontal) radiators each with the phase of the feed and with the spacing adjusted according to the directivity required.

The development of this method is largely due to Dr. G. H. Brown (USA) who was incidentally responsible for the ground plane antenna more or less as we know it today. In 1937 G. H. Brown published his now classic set of radiation patterns (Fig. 3.1) which shows some of the possible combinations of directivity with end-fire and broadside arrays. In the end-fire case, radiation is generally in line with the two radiators, whereas for the broadside array it is at right angles to the line of the array. For example, the radiation from a typical end-fire configuration of two verticals spaced $^1/_8\lambda$ and phased 180°, i.e. in phase opposition, can be seen in Fig. 3.1 and marked x. A typical broadside pattern is the one marked y where the spacing is $^1/_2\lambda$ and the elements are driven in phase. Spacings greater than this are little used for amateur radio but nevertheless offer design possibilities for small beams with higher gain.

A typical configuration based on this idea is shown in Fig. 3.2 in which two half-wave centre fed radiators are spaced $D = {}^1/_8\lambda$ apart and driven 180° out of phase. Such an array can be used vertically as shown, or horizontally, depending on the polarisation required. When used vertically (looking down on the ends of the radiators) the field pattern will be as in (a) and the vertical angle radiation, i.e. on a plane parallel to ground, will be as in (b). It

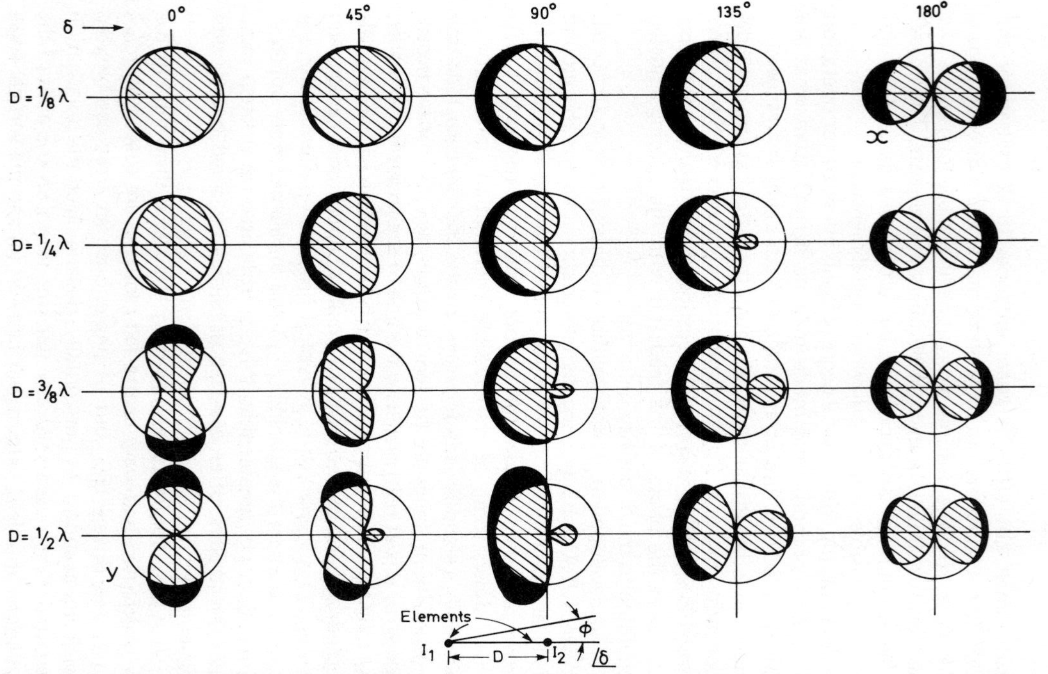

Fig. 3.1. *Radiation patterns of two vertical elements as a function of phase difference and distance apart.*

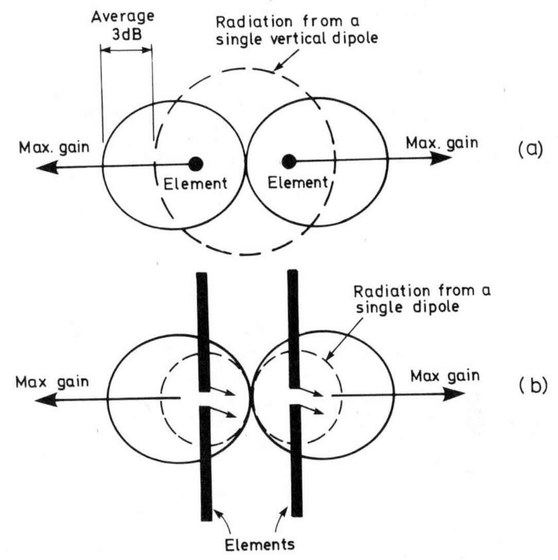

Fig. 3.2. Typical field patterns of two half-wave elements, centre fed, spaced $^1/_8\,\lambda$ apart and driven 180° out of phase

will be appreciated therefore that with this arrangement the radiation pattern would be approximately the same whether the antenna were operated vertically or horizontally. However, radiation in two directions is not always desirable despite the gain achieved over a dipole. By keeping the spacing between elements the same, i.e. $^1/_8\,\lambda$ and by feeding one element 135° out of phase with the other, it is possible to produce a unidirectional antenna with gain that can be used horizontally having one forward pattern as in Fig. 3.3(a) or vertically, still with one forward pattern, although cardioid as in Fig. 3.3(b). The forward gain in either mode is about 3.9 dB.

An antenna known as the HB9CV is based on this arrangement and employs a gamma matching system to enable a direct feed to be made with 50 or 70 Ω coaxial cable. The claimed gain is about 5 dB and whilst the antenna is compact many who have constructed it have reported difficulty in obtaining a low VSWR. Details of an antenna of this design are given in Fig. 3.4.

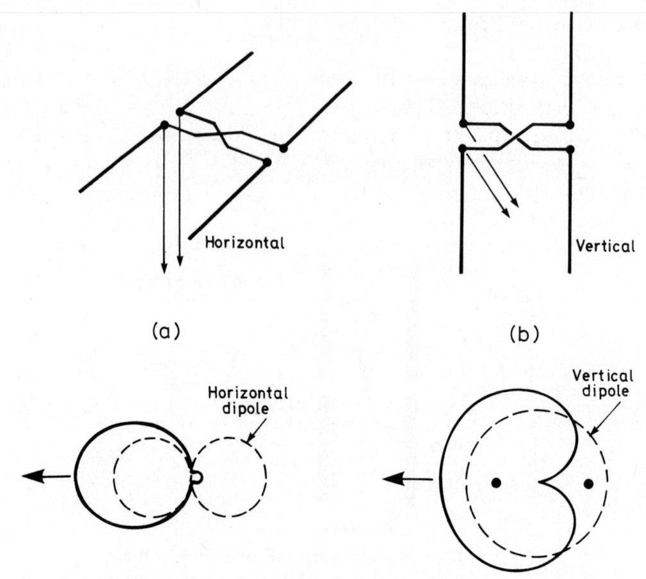

Fig. 3.3. Configuration and field patterns of an end fire array with two elements spaced $^1/_8\,\lambda$ apart and fed 135° out of phase

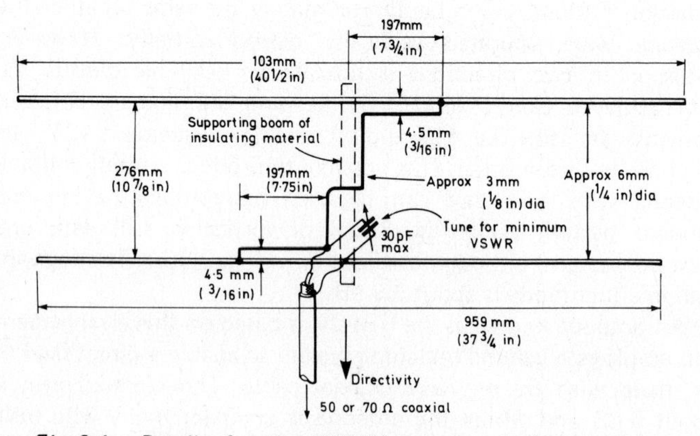

Fig. 3.4. Details for the HB9CV gamma matched compact beam antenna

The ZL Special series

A system also based on the end-fire array with 135° phase shift in the feed to one element was developed by the author some 26 years ago for the h.f. bands and called the ZL Special. As the front and rear elements are of different length, this antenna also exhibits a reflector – director action with a resultant gain of about 6 dB over a dipole. It can be fed directly from 50 or 70 Ω

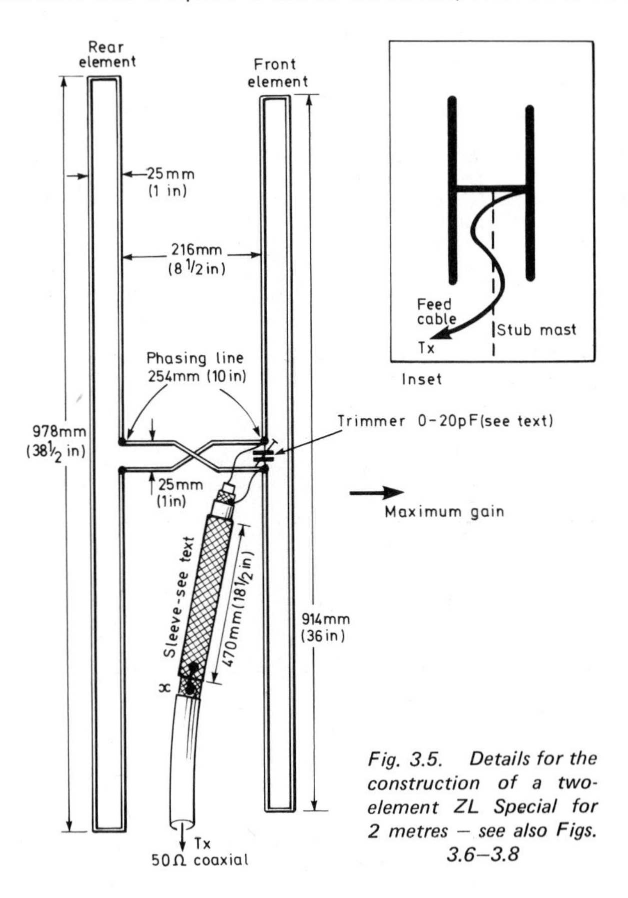

Fig. 3.5. Details for the construction of a two-element ZL Special for 2 metres – see also Figs. 3.6–3.8

coaxial cable, and when properly adjusted a VSWR of much less than 1.5 to 1 is possible.

A ZL Special for 2 metres

This is not a difficult antenna to construct, and the elements and phasing line may be made from copper wire, coat hanger or galvanised iron wire, copper or aluminium tube, or even 300 Ω ribbon feeder. Quite a large number of 2 metre versions have been

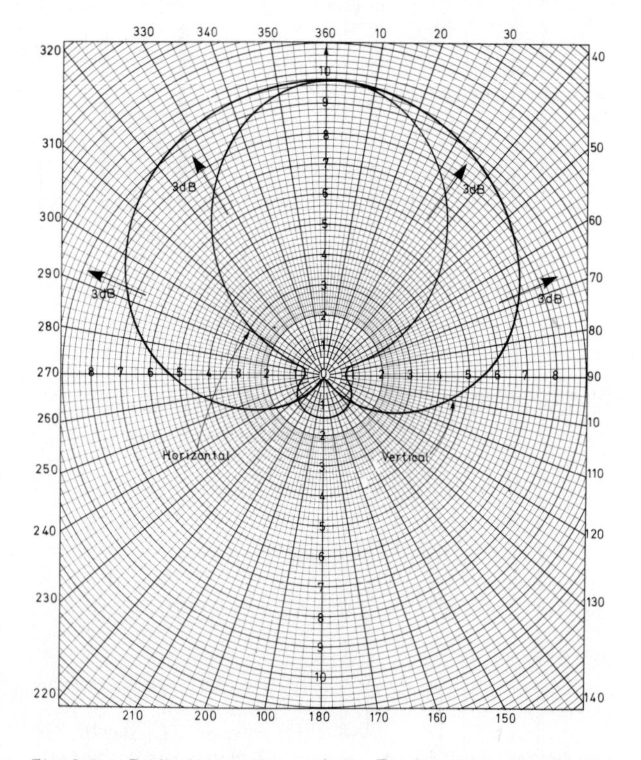

Fig. 3.6. Radiation patterns of the ZL Special in vertical and horizontal mode

made and tested and being a very compact aerial it has also proved ideal for use indoors. It has a broad band characteristic and properly constructed and tuned should provide a VSWR of less than 1.5 to 1 over the whole 2 metre band. Constructional and other details are given in Figs. 3.5–3.7 and provided the element spacing and phasing line length is maintained the materials for these may be as mentioned above. The small air-spaced variable

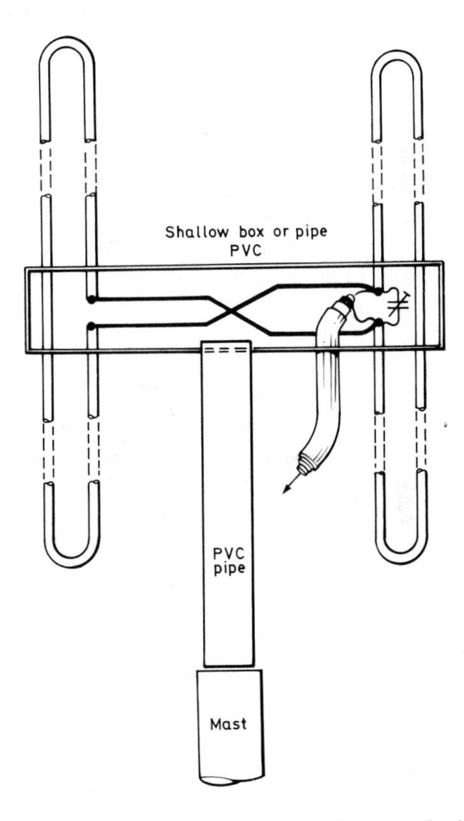

Fig. 3.7. Housing the phasing line and feed connections etc. of the ZL Special

capacitor (max. capacity 20 pF) is necessary to achieve a good match with 50 or 70 Ω coaxial cable.

This capacitor is adjusted for minimum VSWR with the full length of coaxial cable to be used already attached to the aerial. When the antenna is operated vertically it is important to curve the feed cable away from it and downwards as shown in Fig. 3.5 (inset). The stub mast supporting the antenna must be of non-conducting material, e.g. wooden dowel or plastics PVC pipe, and extended to at least a few inches below the extremities of the elements. The radiation patterns in both vertical and horizontal modes of operation are shown in Fig. 3.6. Note the balance to unbalanced feed sleeve over the end of the coaxial cable where it joins the feed point. This is 470 mm ($18\frac{1}{2}$ in) long and bonded to the main cable screening braid only where shown at X. The end nearest the feed point is not connected anywhere and must not come into contact with the feed point connections. For use outdoors, the feed point connections, the small capacitor and the end of the coaxial cable must be protected against rainwater. A small plastics box with a tight fitting lid can be used for this purpose as outlined in Fig. 3.7. A completely weatherproof version of the ZL Special used by the author on a small cabin cruiser (G2BCX/MM) and at a height of only 3 m (9 ft) above water level has proved exceptionally good with much continental and UK DX to its credit. The elements of this weatherproof version are completely enclosed in flat section (electric cable) plastics tube as in the photograph in Fig. 3.8. As this antenna will go into the boot of a car it is ideal for portable operation with a suitable sectioned mast.

Simple vertical end-fire array

This is a vertically polarised antenna that is relatively easy to construct and that will provide a gain of about 4 dB in two directions. It needs to be turned only through 90° to achieve all round coverage and being compact could prove useful to flat dwellers unable to put up an outside aerial. It consists of two half-wave vertical radiators end fed 180° out of phase to produce

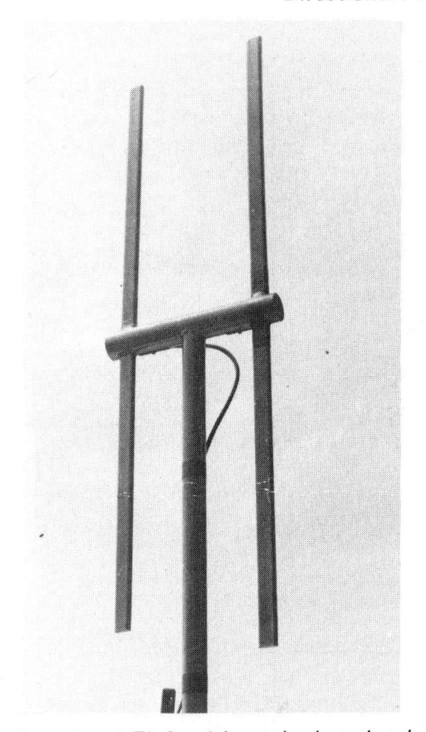

Fig. 3.8. A ZL Special completely enclosed in plastics tube for complete protection against weather (G2BCX/MM)

a radiation pattern as in Fig. 3.9. It may be assembled on a light wooden frame with the elements made from coat hanger wire, or better copper wire of 14 or 16 swg (2.03 or 1.63 mm diameter), and then set up to rotate on a floor stand as the diagram suggests, thus making it suitable for indoor use. Adjustment for minimum SWR is made by sliding the shorting bar on the stub in conjunction with moving the tapping points of the 50 Ω feeder. The approximate positions of these are given in the diagram but very little further adjustment will be needed to obtain a VSWR approaching 1 to 1. The radiation pattern is shown in the inset of Fig. 3.9.

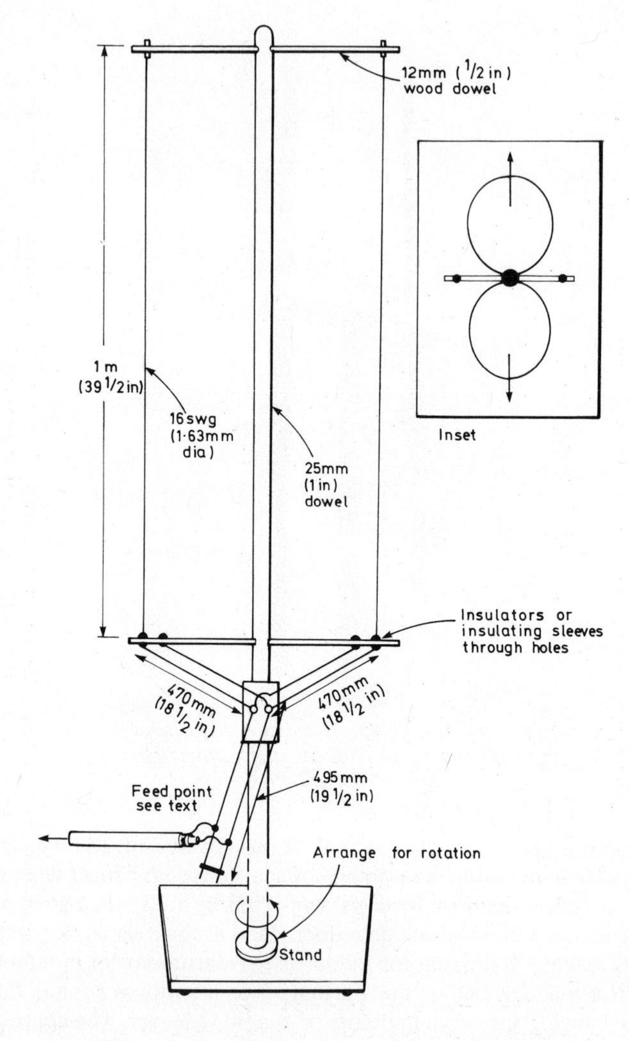

Fig. 3.9. Details for the construction of a simple bi-directional end-fire array for indoor use. Arranged for rotation through about 180°

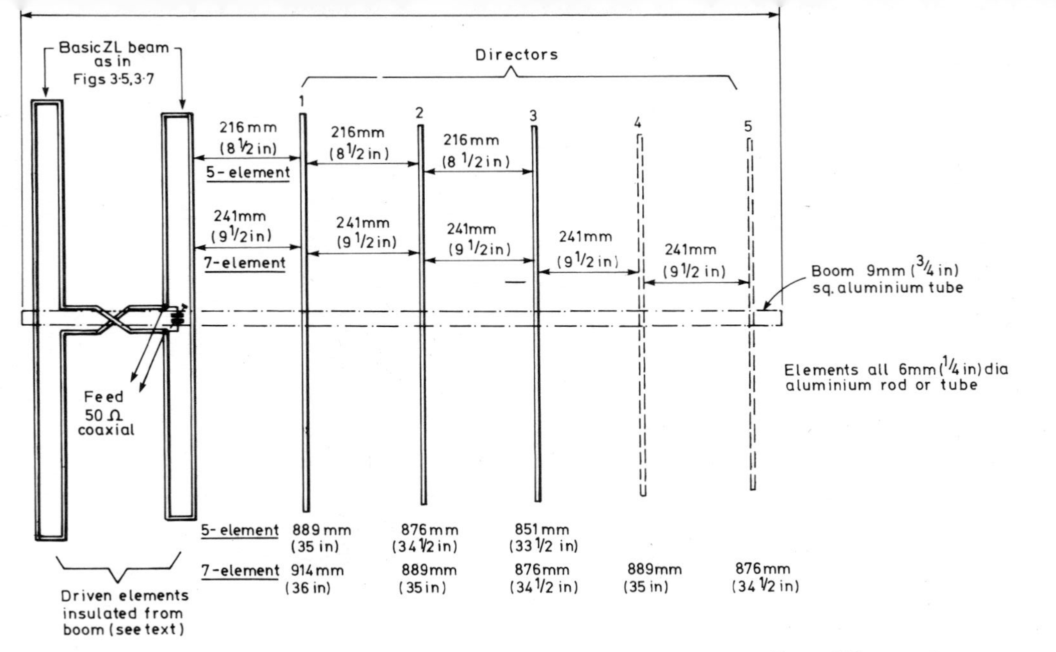

Fig. 3.10. Details for constructing a five- or seven-element ZL Special beam. Note differences between element spacing and lengths for the five- and seven-element version. See also Figs. 3.5 and 3.7 for construction of driven elements

The ZL Special multi-element arrays

Since the basic ZL special is a driven array and the radiation pattern is unidirectional, it has been found possible to add parasitic elements to increase the forward gain. No reflector is required because this is already integral with the ZL special, so only directors are used. One director mounted 0.12 λ in front of the driven element will increase the forward gain by about 1.5 dB, thus providing a total gain of 7 to 7.5 dB, which is quite considerable for a beam antenna measuring only 510 mm (20 in) from front to rear.

The addition of three directors, making the ZL a five-element beam and still only about 1 m (40 in) long, will yield a forward gain of a little over 9 dB which is nearly comparable with that from an eight-element Yagi with a physical length of around 2.8 m (110 in), i.e. over twice the length of the five-element ZL.

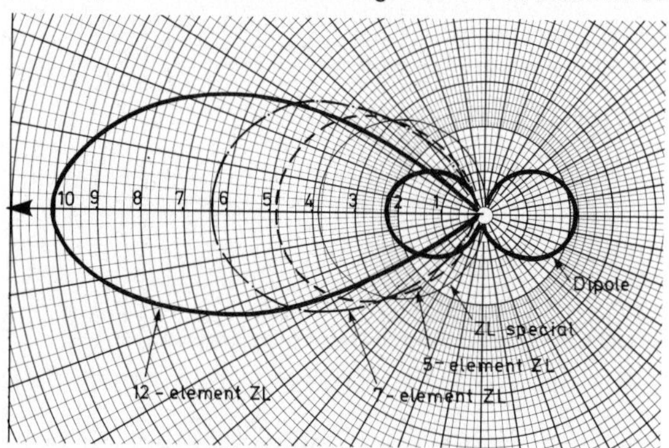

Fig. 3.11. Radiation patterns of the ZL series of antennas described in this chapter

The construction of the driven element section for a five- or seven-element ZL beam is the same as for the basic antenna shown in Figs. 3.5–3.7, but this section must be insulated from the boom which will also support the directors and which is made from 18 mm (¾ in) square aluminium tube to a length determined by the number of directors to be used. Details for director lengths and spacing, according to whether a five- or seven-element version is to be made, are given in Fig. 3.10. The directors are screwed to the boom (or fixed into holes bored through) at their exact centres. The phasing lines linking the driven elements may be 16 or 14 swg (1.63 or 2.03 mm diameter) copper wire and to ensure they do not touch, insulating sleeves should be fitted.

The relative gains and radiation patterns of the ZL series of antennas are shown for comparison in Fig. 3.11. The ZL Special two-element has a gain of about 6 dBd, the five-element about 9.5 dBd, and the seven-element about 10.5 dBd. To achieve these maximum gain figures it is important that the antennas are correctly adjusted, that good quality low loss coaxial cable is used, and that the VSWR is not more than about 1.2 to 1.

A 12-element ZL beam

This antenna is the result of extensive development and if carefully constructed will yield a gain of at least 13.5 dBd with a VSWR across the band of not more than about 1.5 to 1. Good quality low loss coaxial cable is also very important. Even though a low VSWR may be obtained, the loss on inferior cable can be quite high resulting in poor performance from the antenna.

Construction of the 12-element version is shown in Fig. 3.12 which contains details of the boom length and the lengths and spacing of the directors. There is a difference however in the feed system and phasing line and also in the construction of the elements. This should all be clear from Figs. 3.13 and 3.14. The driven elements are turned so they are at right angles to the antenna, and not as shown for the other ZL antennas. The phasing line itself is made from 300 Ω ribbon feeder (267 mm ($10\frac{1}{2}$ in) long) and note the addition of the small stub across the feed point of the rear element and the phasing line.

Also across the feed point on the front element is a 114 mm ($4\frac{1}{2}$ in) length of 50 Ω coaxial cable (open at one end) which acts as a low value tuning capacitance. To obtain the lowest VSWR across the band the length of this piece of coaxial may need adjustment but do this at 145 MHz. Fig. 3.13(b) illustrates how

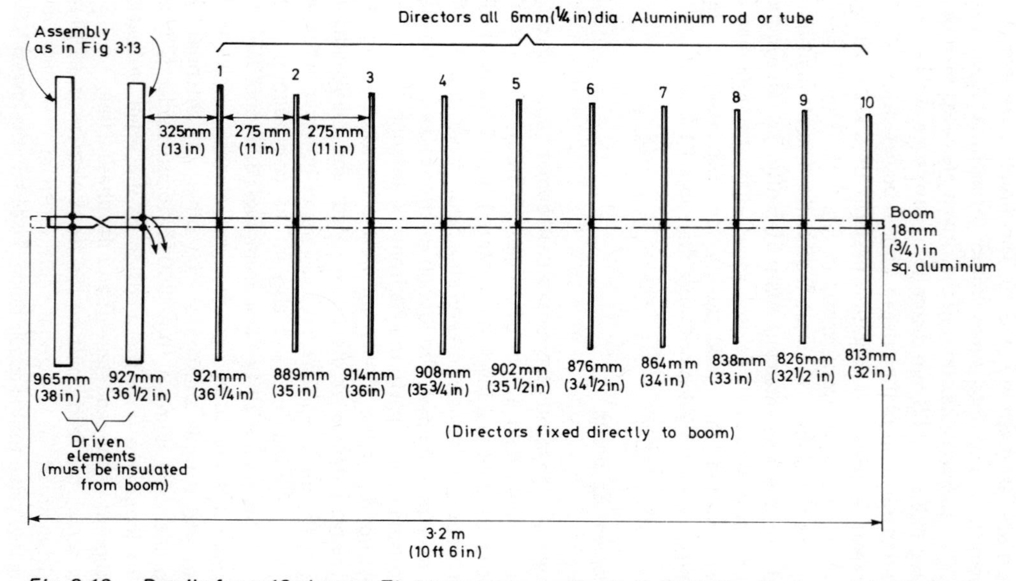

Fig. 3.12. Details for a 12-element ZL beam antenna with about 13.5 dB gain over a dipole (see also Figs. 3.13 and 3.14)

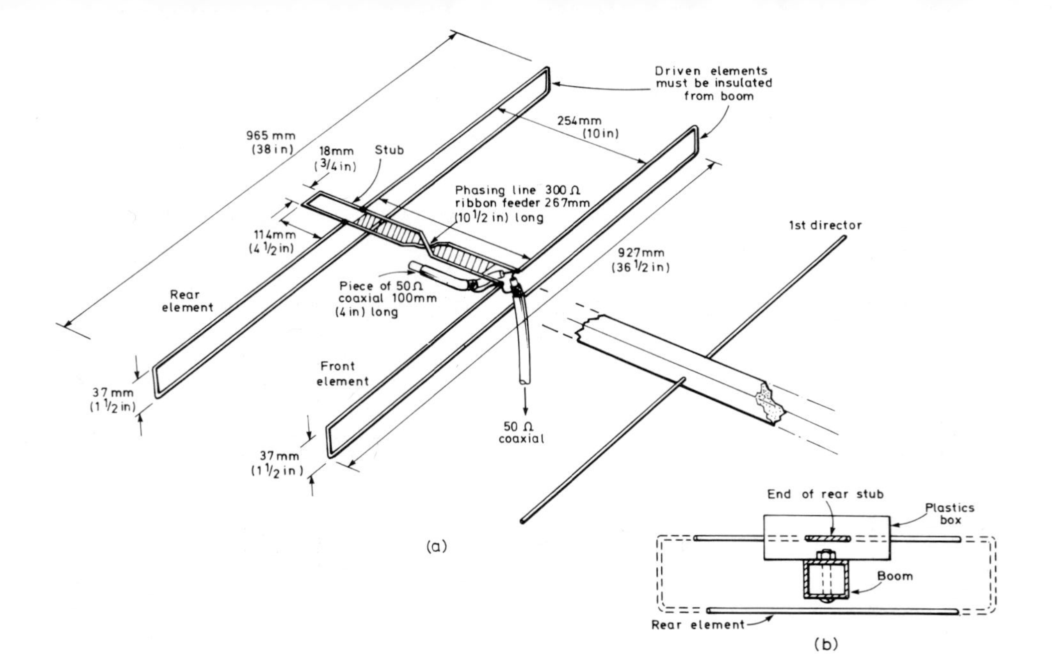

Fig. 3.13. Arrangement of the driven elements phasing line and tuning stubs of the 12-element ZL beam

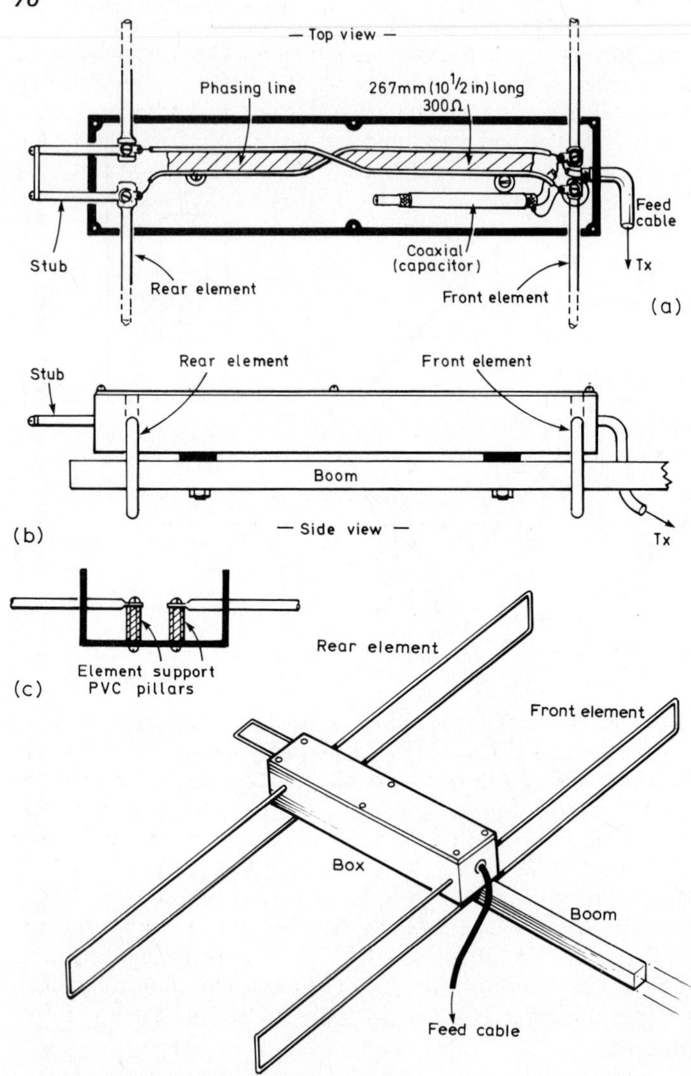

Fig. 3.14. Method of protecting the phasing line and cable connections and isolating the driven elements from the boom

the elements are arranged to run underneath the boom but not in contact with it and how they are supported by the insulating box. This box must be long enough as in Fig. 3.14(c) to enclose the phasing line and feed points etc. and can be made from hard wood or plastics. If wood is used, make sure it is dry and after

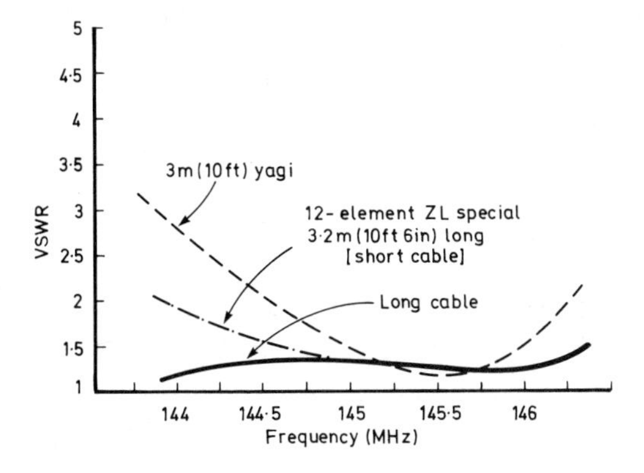

Fig. 3.15. VSWR v frequency graph of 12-element ZL beam 3.2 m (10 ft 6 in) long and a Yagi array of 3 m (10 ft) length (see text)

assembly and adjustment apply at least three coats of water resistant varnish. Use a sealant around the edges to make sure that water cannot seep through the lid. The radiation pattern for the 12-element version is shown in Fig. 3.11. The curves in Fig. 3.15 show the VSWR obtainable over the 145 MHz band (144 to 146 MHz) with a 12-element ZL Special. With a long cable (more than 16 m (50 ft)), the curve will tend to flatten out towards the lower frequencies. The otherwise shallow curve indicates a wide bandwidth. The VSWR v frequency graph of a typical long Yagi is shown by comparison. This is deep, denoting sharper resonance but a much higher reflective power when off resonance.

A photograph of the finished 12-element ZL beam (Fig. 3.16) as designed and used by the author is shown mounted on its rotator. It is also equipped with an electrically driven vertical to horizontal rotator which is the small oblong box at the centre. The antenna on the stub mast above the ZL beam is a Slim Jim.

Horizontal to vertical rotator

A small rotator to orientate a beam antenna from vertical to horizontal mode is not difficult to make and does not have to be of massive construction. All that is really required is a small reversible direction (d.c.) motor and a train of gears to develop enough power to turn an arm supporting the beam through 90°. The idea is outlined in Fig. 3.17(a) and (b). The motor is a small

Fig. 3.16. The author's original 12-element ZL beam complete with 360° rotator and horizontal/vertical rotator. Vertical antenna above the beam is a Slim Jim

12 V type and the gears are made by Meccano. The micro-switches are operated by cams on the main spindle so as to cut the d.c. voltage to the motors at the end of either direction of run, i.e. at zero and/or 90°. A suggested circuit for the auto-stop and motor reverse is given in Fig. 3.18.

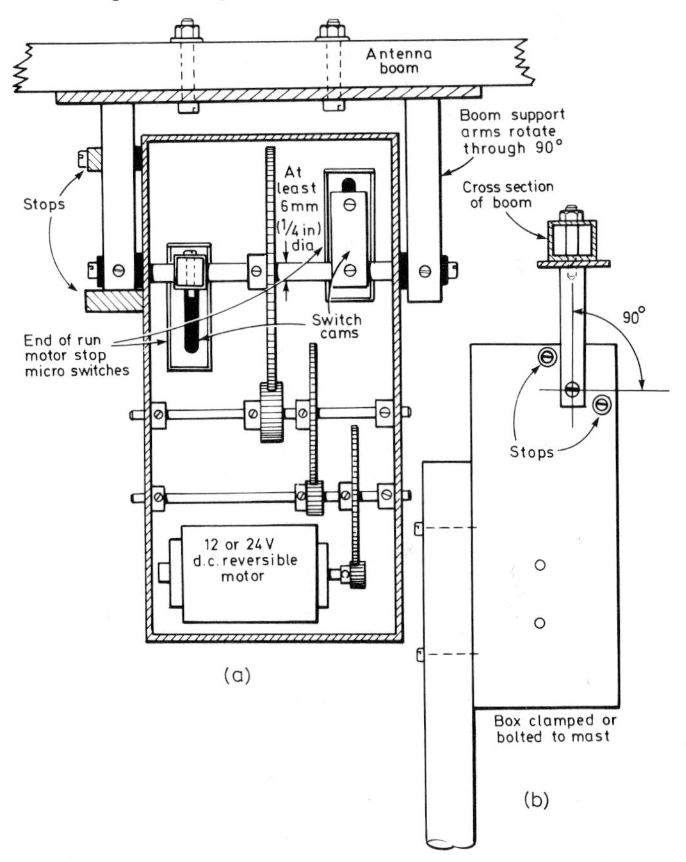

Fig. 3.17. An idea for a vertical to horizontal beam rotator. The cams are used to operate micro-switches that cut power to the motor at end of traverse. Box – Eddystone diecast or metal box with lid. Walls must be thick enough to provide support for bearings and gear wheel spindles

Parasitic arrays

The basic principle of a parasitic array was explained briefly in Chapter 1 although, as with the ZL series of beam antennas, one or more of the elements may be active. Generally speaking a parasitic element obtains its power from an active or driven element by electromagnetic coupling, but at the same time it can re-radiate this power which may be quite considerable. Since the parasitic element or elements may be in very close proximity to the driving element(s) and to each other, power may also be transferred from one parasitic element to another. The behaviour of driven and parasitic antenna elements in close proximity is quite complex, as such factors as phase relationship and impedance

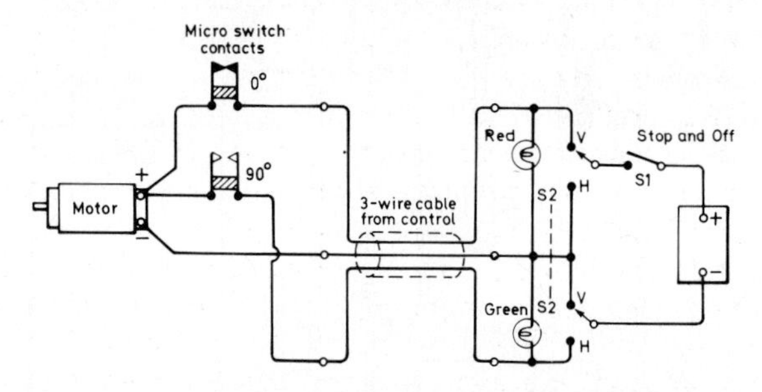

Fig. 3.18. Suggested circuit for controlling the vertical/ horizontal rotator. Appropriate contact is closed when motor is running. As in circuit the zero degree micro-switch is closed and when the beam reaches vertical this contact opens thus cutting power to the motor. Reversing the switch S2 to horizontal will now turn beam to horizontal. Micro-switch for 90° will be closed for this but will open when beam reaches full horizontal. The stop and/or off switch S1 can be used to stop the rotation at any point, e.g. for slant operation of the beam at any angle between 0 and 90°

are involved. First however, let us take a pair of dipoles very close together. When power is applied to one dipole and current flows, a voltage and therefore a current will be induced into the other dipole from the electromagnetic field created around the driven dipole. The current now generated in the second dipole will in turn create a magnetic field around itself and induce a voltage in the first (driven) dipole, the total current in this, without at the moment taking phase into account, being the sum of the original current in the first dipole and the re-induced current from the 'parasitic' dipole. The process of this action however, causes a change in the impedance of the original driven dipole because of mutual coupling, and this creates a mutual impedance between the driven and parasitic elements. The final actual impedance of the driven element is the sum of its self impedance and its mutual impedance with other elements in close proximity.

When the separation of two elements is a fairly large fraction of a wavelength however, a period of time passes before the radiation from the driven element reaches the next (parasitic) and is returned to the original driven element. Because of this there will be a phase difference between the original current (in the first element) and the current induced into it by the parasitic. This phase difference will depend on the amount of 'spacing' between the two elements and therefore the original and induced currents can be almost completely in or out of phase. In the first case the combined current is large and the impedance of the driven element is small, and in the second case the current is small and the impedance therefore large. For intermediate spacings and phase relationships the impedance of the driving element will be greater or smaller to a given extent. Another factor that affects impedance is the tuning of the parasitic element. If this is not exactly resonant, the current that flows in it will either lead or lag the phase that would have existed had the element been exactly resonant. This, in turn causes a phase advance or delay that changes the phase of the current re-induced into the driven element. In other words the magnitude of current in the parasitic element and its phase relation to the current in the driven element depends on tuning as well as spacing. The parasitic element may have a fixed length of $^1/_2\lambda$, the tuning being accomplished by inserting a lumped reactance in series with the element at its centre point. Alternatively the parasitic element may be continuous and the tuning accomplished by altering its length. This method is more simple and practical and as a result, the parasitic *directors* in Yagi type arrays as they are often called, are made shorter as the distance from the driving element increases. On the other hand the tuning of a parasitic can be such as to make the element electrically longer by adding reactance, or more practicable, the element can be made physically larger to achieve the same result and this enables the element to be used as a *reflector*. With this arrangement the phasing of the fields between driven element and reflector are such that little or no radiation occurs to the rear of the reflector but its forward field adds to the forward field of the driven elements.

Whether a parasitic element operates as a reflector or director is determined mainly by the relative phases of the currents in the driven and parasitic elements. With an element spacing of a $\frac{1}{4}\lambda$ or less the current in the parasitic element will be correctly phased if its tuning is adjusted to be on the low frequency side of resonance (inductive reactance). The parasitic will then act as a reflector. If the tuning is such that resonance is on the high frequency side (capacitive reactance), then a parasitic will behave as a director. The spacing between the driven element and its reflector and/or director largely determines the forward gain that can be attained.

The basic 2-element parasitic array

The maximum theoretical gain that can be obtained with one driven element and a single parasitic is generally a function of tuning and spacing. The result of an analysis by G. H. Brown is given in Fig. 3.19 in which the two curves show the gain that can be obtained with different spacings when the parasitic element is tuned to operate either as a reflector or director. If a parasitic element is tuned to operate as a director, the highest gain is attained when the spacing is about 0.1λ, as in Fig. 3.20. With the parasitic tuned as a reflector the optimum spacing for highest gain is about 0.15λ. These figures assume no losses and that the elements are also tuned for maximum gain. In practice the gain achieved is somewhat less.

With parasitic arrays the radiation resistance at the centre of the driven element varies with the spacing and tuning of the parasitic elements, and with spacings around 0.1λ may be quite

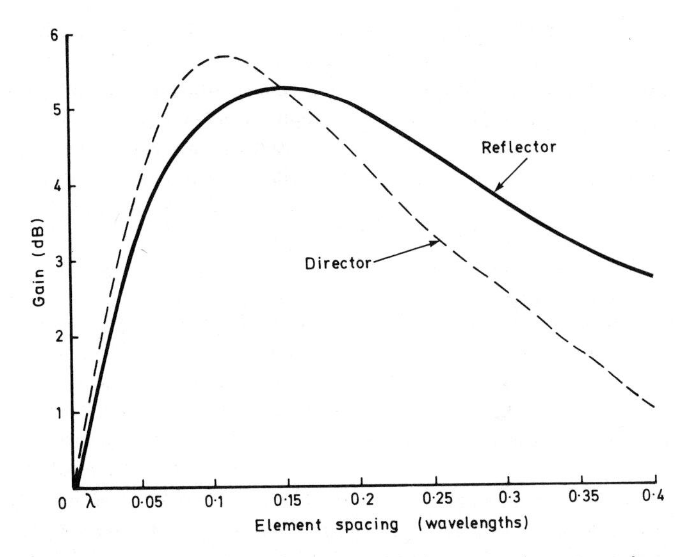

Fig. 3.19. Curves showing relationship between element spacing and gain with one driven and one parasitic element

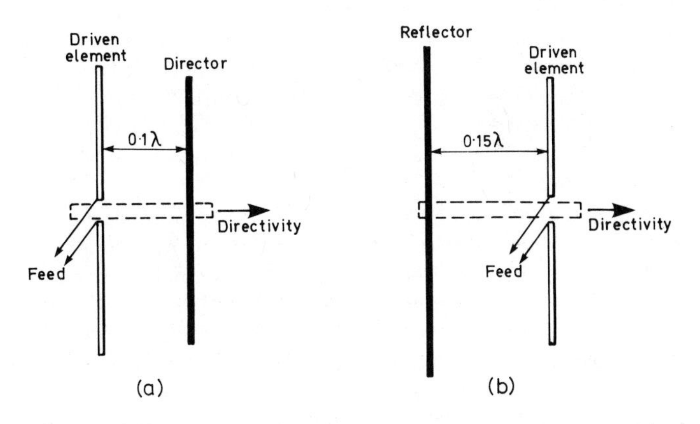

Fig. 3.20. Configuration of driven and parasitic element used as either director or reflector

low, e.g. in the region of 14 Ω. This does, however, tend to reduce efficiency as power fed to the antenna is also dissipated in heat rather than radiated, due to the radiation resistance becoming comparable with the ohmic loss resistance of the system. The loss resistance can be reduced by employing low resistance conductors which means that they must be of fairly large diameter copper or aluminium tubing, especially for antennas for operation at VHF and UHF. Fortunately antennas for these frequencies are physically small and the elements may be self supporting. Any insulating material used in construction, and particularly at points where voltage maxima occur, must be of the highest quality otherwise losses can occur.

Construction of parasitic arrays

The variety of parasitic or Yagi type arrays is too great to include details for the construction of all of them in this book; references given at the end of this chapter will provide further information. Yagi type arrays can consist of combinations ranging from a basic driven element and parasitic reflector, to driven element, reflector and up to 20 or more directors. Moreover, such arrays can be stacked one above the other and/or arranged in bays side by side and also stacked. For example there are commercially available arrays consisting of eight separate 10-element Yagis stacked and bayed with a gain approaching 25 dB, which is a forward power gain of over 300 times, or an ERP from the antenna of 300 times the power fed into it.

The design of Yagi arrays and the ultimate power gain, bandwidth and beam width is largely dependent on the overall length of the array, the total number of elements employed and their lengths and spacing. Arrays with a large number of elements (the total number includes the driven element or elements) are usually called long Yagis, and these are fairly practical at VHF and UHF. Such arrays can be combined to provide both vertical and horizontal polarisation and really consist of a duplication of an otherwise single array with one horizontal and one vertical. These can be fed in such a way as to produce fully vertical or horizontal polarisation, or slant, elliptical or circular polarisation. An alternative method of changing from horizontal to vertical is to use a single array with a motor drive to physically orientate the array through 90°, as described earlier in this chapter.

There is little point in constructing parasitic arrays with less than five or six elements for operation at VHF and UHF as the gain for a simple 2-element system is hardly worth while when so much more can be attained with the addition of a few extra directors. One of the problems with Yagi type arrays, however, is obtaining a reasonably good match with 50 Ω cable, or indeed 70 Ω cable, as the impedance of the driven element, be this a dipole, becomes very low in the presence of a reflector and directors. For 70 Ω feed a folded dipole may be used as the driven element as its self impedance of approximately 300 Ω is reduced to about 70 Ω when in close proximity to a reflector and director. Probably the most common method of matching a 50 Ω cable to a parasitic array with a normal dipole driving

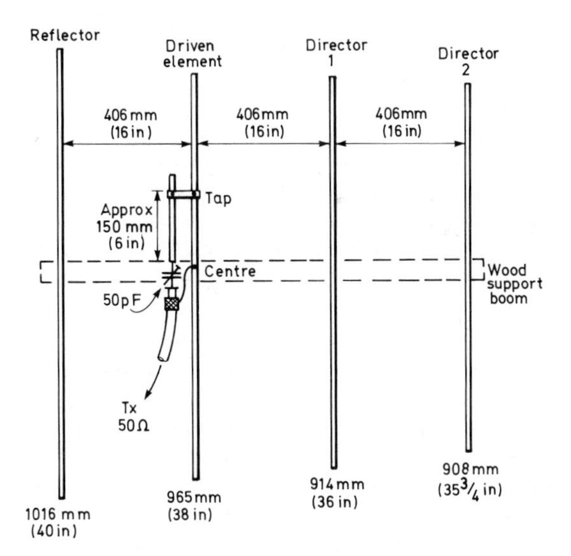

Fig. 3.21. Dimensions for the construction of a gamma matched four-element Yagi

element is to use a gamma match as in the arrangement for the 4-element simple array shown in Fig. 3.21. The support boom could be metal as the centres of all the elements are at zero r.f. potential. The elements may be aluminium or copper tubing of 9–12 mm ($^3/_8$–$^1/_2$ in) diameter, as also is the gamma stub at the feed point. The 50 pF capacitor should be a small air spaced type, and this and the tapping point of the gamma stub are adjusted for minimum VSWR. Some form of protection will be necessary for the feed point and capacitor, and for this a small plastics box with tight fitting lid could be used to prevent moisture getting to the tuning capacitor and the end of the coaxial cable. It is always most important to prevent rainwater getting into the open end of a coaxial cable for if this happens not only will the performance of the antenna be degraded but the cable itself ruined.

A 7- to 10-element Yagi (Fig. 3.22)

This array employs a folded dipole with a ratio match (see also Chapter 4), and the boom which must be of hard dry wood is 3 m (10 ft) long. The elements may be supported through holes in the boom which needs protection from weather with two or three coats of paint or varnish. After final assembly the whole antenna can be given one or two coats of varnish. The cable connections to the feed point must also be protected against rainwater, and for this a small plastics box screwed to the boom could be used. The reflector and director elements may be made from copper or aluminium tube of 6–12 mm ($^1/_4$–$^1/_2$ in) diameter so they are self supporting. The ratio folded dipole is constructed as shown in Fig. 3.22(b) and can be fed directly with a 50 Ω coaxial cable and balun loop. The dimensions given for the driven end parasitic elements should provide optimum performance at band centre, i.e. 145 MHz. Additional directors can be used to increase forward gain and these are shown dotted in Fig. 3.22(c). It is important to use the correct diameter material for the driven element sections, one thick and one thin, as the diameter ratio determines the impedance at the feed point. The balun $^1/_2\lambda$ section may be of the same coaxial cable used to couple the

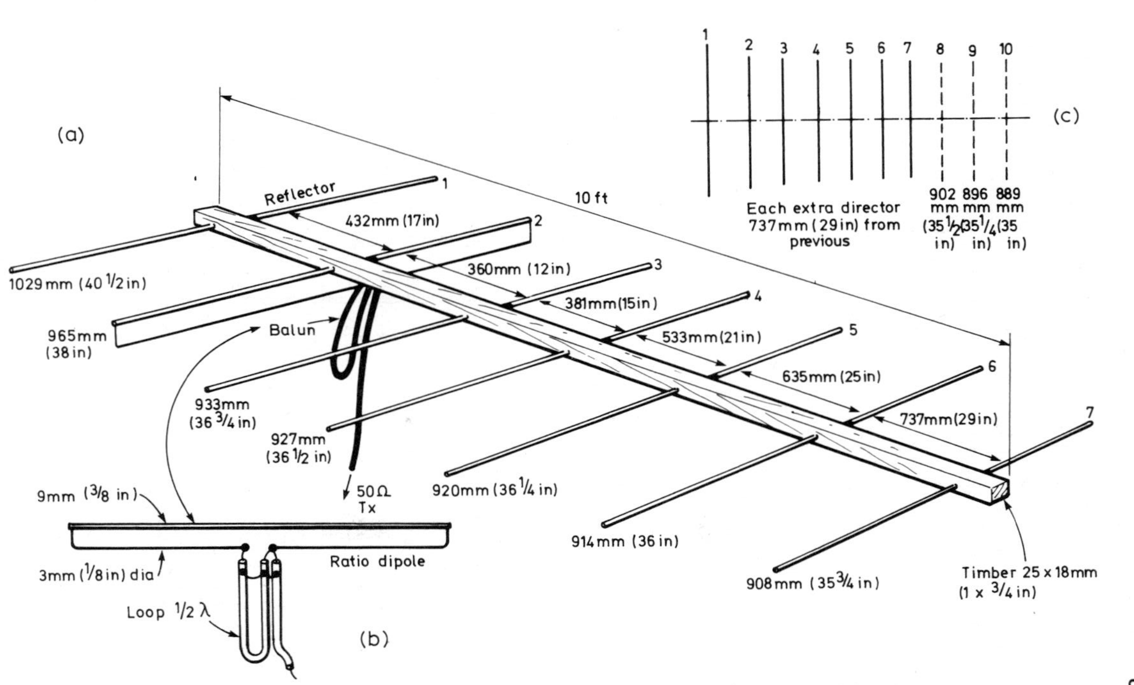

Fig. 3.22. Details for the construction of a seven- or ten-element Yagi array with ratio matching

antenna to the transmitter and will be approximately 686 mm (27 in) long if the velocity of the cable is 0.66. The forward gain of this antenna should be in the region of 11 dBd with seven elements and 12 to 13 dBd with ten elements, which will make the total length of the antenna nearly 4.6 m (15 ft). More gain than this can be obtained with a 12-element ZL beam which is only 3 m (10ft 6 in) in overall length.

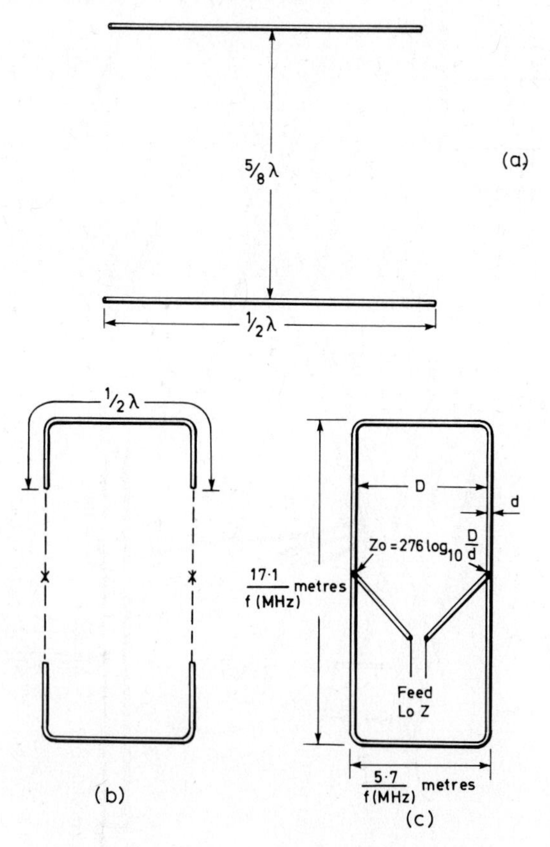

Fig. 3.23. Formation of the skeleton slot radiator – see text for explanation

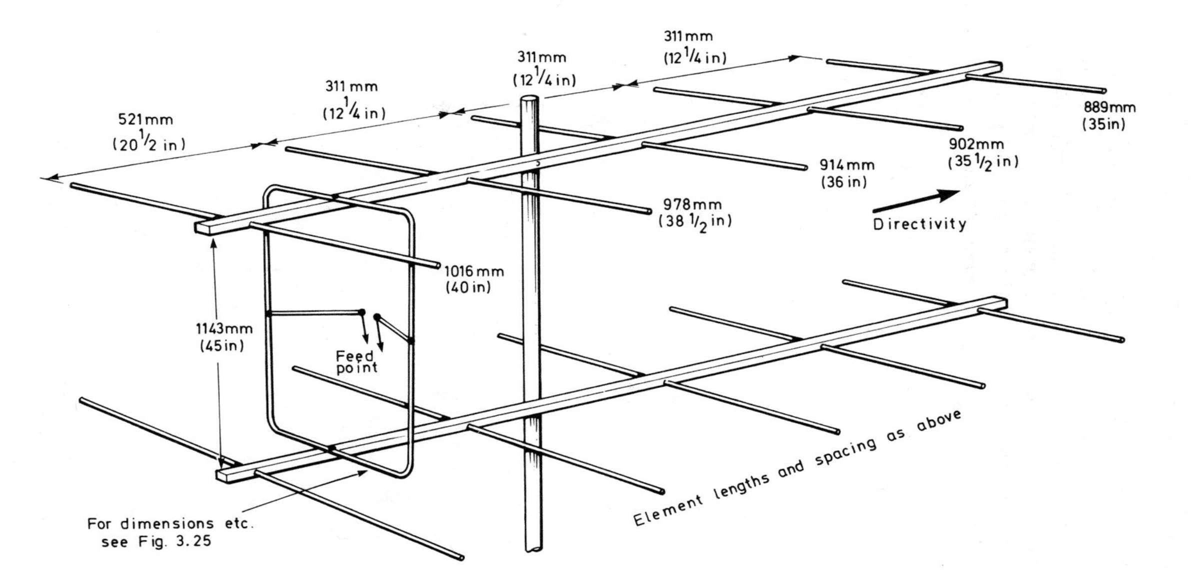

Fig. 3.24. ***Details for the construction of a 6 over 6 skeleton slot driven beam antenna***

Stacked skeleton slot array

The 'skeleton' slot system was developed by B. Sykes G2HCG of J-Beams Limited. The name is derived from the nature of the primary driven element. A normal slot antenna is, as its name suggests, simply a narrow slot of approximately $^1/_2\lambda$ in length cut in a large sheet conductor, which, when driven, radiates as a normal element of wire or tube except that the plane of polarisation

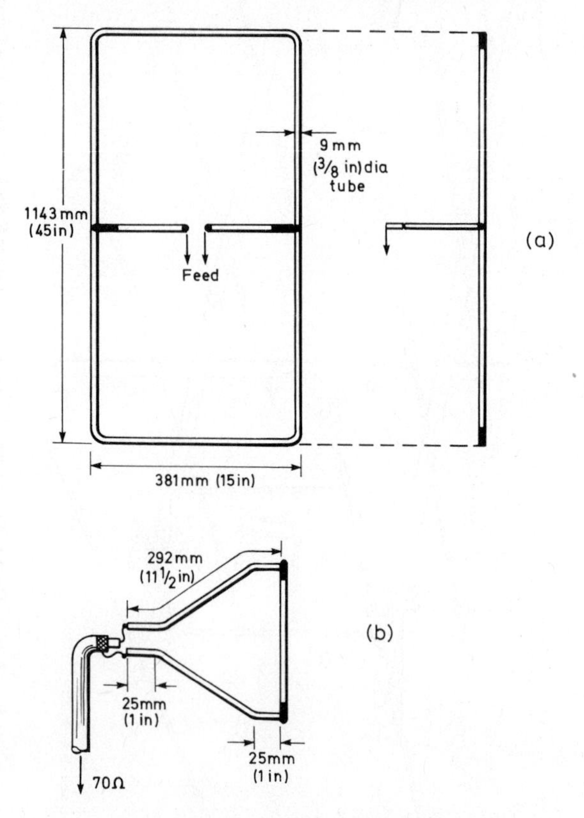

Fig. 3.25. Delta match assembly for driving the 6 over 6 array shown in Fig. 3.24

is 90° opposed, i.e. a vertical slot is horizontally polarised and vice versa. The open or skeleton slot is devised as shown in Fig. 3.23 and could to some extent be regarded as a loop composed of two half-wave radiators (a) folded down as (b) and joined to form the long loop as (c), this could now be considered as a closed length of transmission line with an impedance $Z_0 = 276 \log_{10} (D/d)$ with D as the width of the closed section and d the diameter of the conductor used. To operate with a low impedance feed cable (70 Ω) a form of delta matching is used as in (c) connected to the centre of the loop. Each section of the slot loop, i.e. each of the

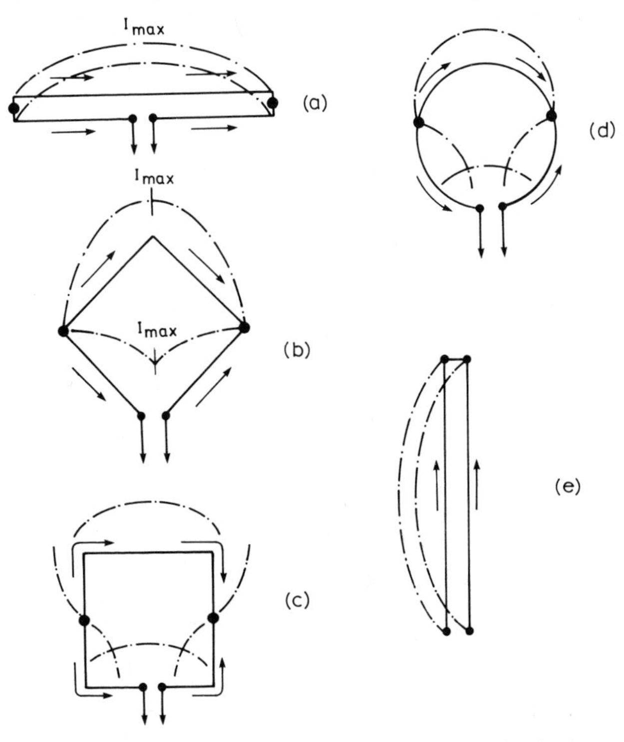

Fig. 3.26. Derivation of square, diamond and circular loop aerials. Arrows show direction of currents and large dots the high impedance (voltage maximum) points

original $^1/_2\lambda$ sections, will now radiate in phase in much the same way as a square loop formed by two $^1/_2\lambda$ radiators. The system is fairly broad band but also has the advantage that it can be used to drive a stacked system of reflectors and directors. Each half of the slot functions as a radiator so the configuration shown in Fig. 3.24 would be regarded as a 'six over six' array and should provide a gain of around 11 dB and a half power beam width of approximately 50°. Details of the slot radiator and matching section are given in Fig. 3.25. Two arrays of this nature are marketed by J-Beams Limited, one being a 'five over five' system with 10.9 dB gain and the other an 'eight over eight' array with 12.6 dB gain.

The quad antenna

The 'quad' antenna may well have been devised by opening a $^1/_2\lambda$ folded dipole as illustrated by Fig. 3.26(a) into a diamond shape as in (b). It may equally have been devised by bending down the ends of two single-element dipoles spaced one above the other by $^1/_4\lambda$ so as to form a square, as in (c), or curved round to form a circle as in (d). This process of re-shaping would finally result in an opened ended section of transmission line $^1/_2\lambda$ long which, if driven via one leg, would become an end-fed folded dipole as in the case of the Slim Jim omnidirectional vertical antenna described in Chapter 2. There is, therefore, nothing magic about a quad antenna, but credit must of course be given to the American radio amateur W9LZX who developed the system in 1939. Square, circular and diamond shaped loops of one wavelength right round make an interesting study in possibilities for arrays with relatively high gain. Loops of this nature have a radiation pattern similar to a dipole, i.e. figure of eight shaped, but also have a gain of about 1.4 dB over a dipole and lend themselves well for the addition of loop or linear reflectors and directors. The Q factor of closed loops of this nature is not too high so they will also operate over a relatively wide frequency band. Gain versus number of elements and length of array etc. has always been a fairly controversial subject, especially when attempts are made to compare the performance of quad arrays with Yagi arrays. By and large there is not much to choose between the two, and the only real problem with a quad array is that construction may be a little more difficult.

A 2-element quad

This is probably the most simple but effective of the vast range of arrays that may be derived from square elements. It has a gain of around 6 dB over a dipole and being fed at the base of one element the arrangement shown in Fig. 3.27 is horizontally polarised. For vertical polarisation it is necessary only to turn the array through 90° so that the feed point is at one side. The spacing between the reflector (larger near element) and the driven element is not critical, and adjustment between 0.15 and 0.25λ will ensure maximum forward gain consistent with minimum VSWR. The feed point must be totally enclosed to prevent water getting to it and the end of the coaxial cable, and a small square electrical plastics junction box could be used. The top ends of the loops may be supported directly by the wood frame as shown as these are maximum current points at virtually zero impedance. The frame may be constructed from 12 or 18 mm ($^1/_2$ or $^3/_4$ in) square batten and given two or three coats of varnish for protection against weather. The mast section used to support the antenna should preferably be of non-conducting material, particularly if the antenna is to be used in vertical polarisation mode. More loop directors may be used to achieve higher gain and these will generally be a few per cent smaller than the driven element with spacing of about 0.15λ in front of the driven element and to each other. Spacing for optimum gain can be carried out with the aid of a field strength meter and with the antenna mounted a few feet above ground in clear surroundings. Elements for quads can be made all the same size, i.e. one full wavelength round and tuned by means of stubs at the centre point with closed stubs for reflectors and open stubs for the directors, as shown in Fig. 3.25. The length of the stubs need not be more than $^1/_8$λ for average ranges of tuning.

The double diamond quad

This is a form of quad antenna developed by the author and is derived from a double folded dipole as shown in Fig. 3.29(a) in which, as the arrows indicate, the currents flowing in each element are in phase. When an antenna of this nature is opened outward as in (b), we have in effect two separate folded dipoles, one

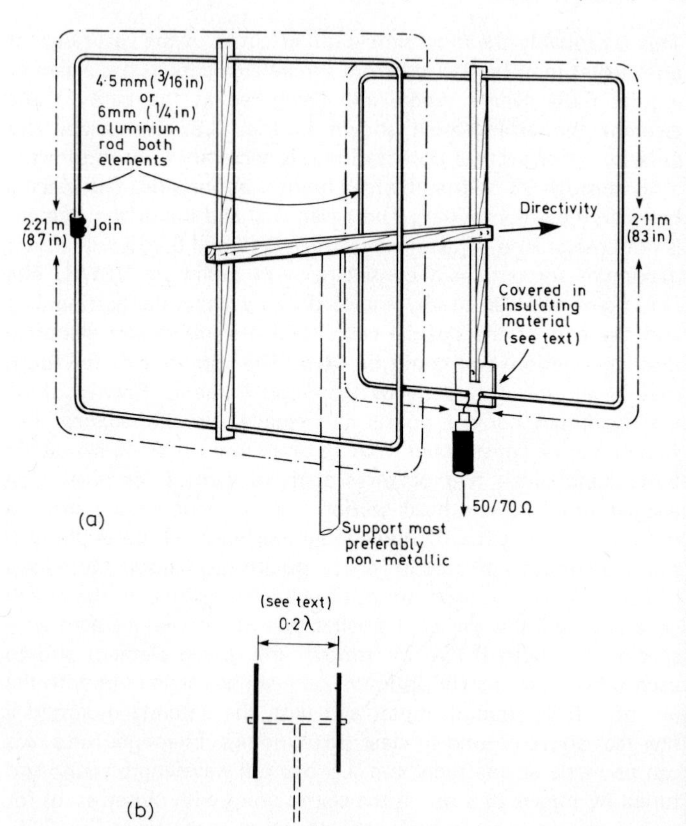

Fig. 3.27. (a) Details for the construction of a simple two-element quad. (b) Spacing between elements (see text)

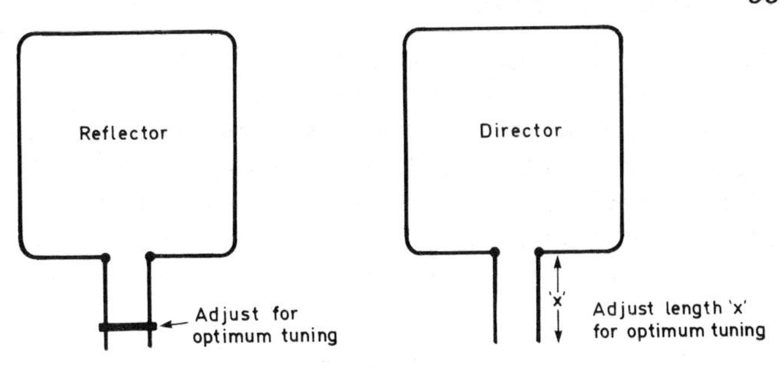

Fig. 3.28. Method of tuning directors or reflectors of same physical size as the driven element in a quad array

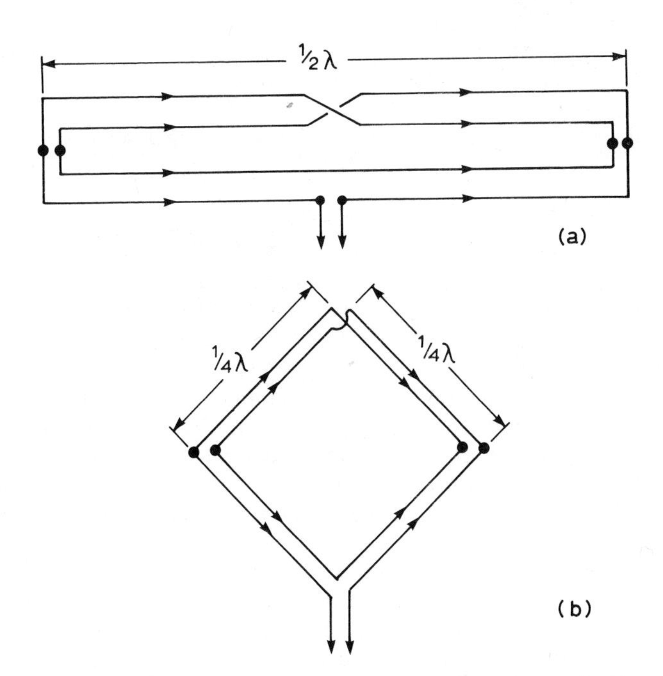

Fig. 3.29. Derivation of the double loop radiator from a double folded dipole

feeding the other at the high impedance points shown by the dots. The current flow remains the same, i.e. in one direction around the double loop and therefore in phase. Such a loop may be fed at a high current, low impedance point as at (b) and by itself will have a gain of around 1.5 dB over a single dipole.

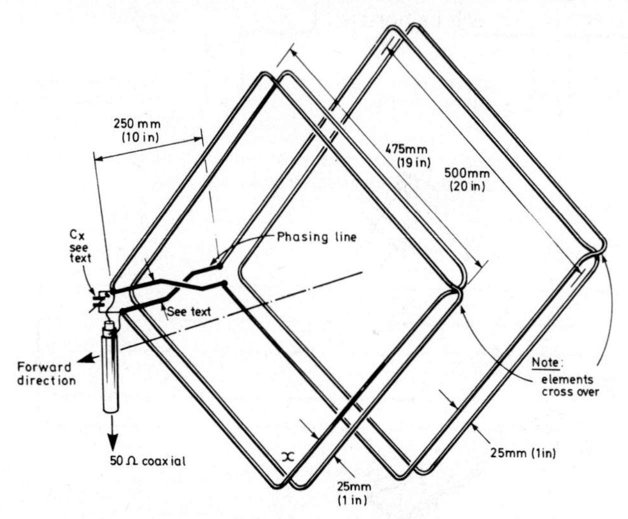

Fig. 3.30. Dimensions of driven elements and phasing line etc. for the double diamond antenna

However, two double loops of this nature like any other pair of radiating elements may be driven together with various degrees of phasing and adjusted and spaced so that one behaves as a driven reflector and the other a driven director in the same way as the ZL Special. Maximum forward gain is achieved when the two double loops are driven so that there is a phase difference of 135° between them, and the space between each is approximately 0.125 λ. The final arrangement is shown in Fig. 3.30, which also includes dimensions for operation on 2 metres. The feed cable may be 50 Ω (coaxial) and the small capacitor C_x of air spaced construction with a value of about 20 pF maximum. It is very important that the double loops cross over as shown. A suggested method of construction is illustrated in Fig. 3.3, for which the support frames are made from 12 to 18 mm ($^1/_2$ to $^3/_4$ in) square batten. With the feed point at one side as in both diagrams the polarisation will be vertical. If however, the antenna is orientated so that the feed point is at the bottom, as shown by X in Fig. 3.30, then polarisation will be horizontal. The only adjustment required

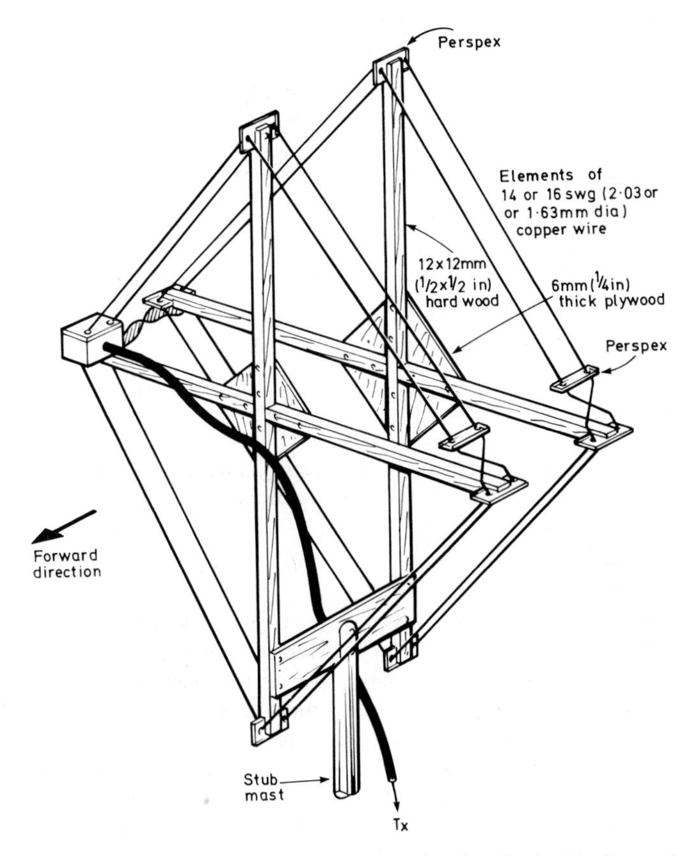

Fig. 3.31. Suggested method of construction for the double diamond quad antenna

is that of the capacitor C_x which should be set to obtain minimum VSWR with the feed cable cut to the length needed by the antenna when finally sited. This antenna is of course directive and the vertical mode radiation pattern is shown in Fig. 3.32, the half power beam width being approximately 52°. The gain over a

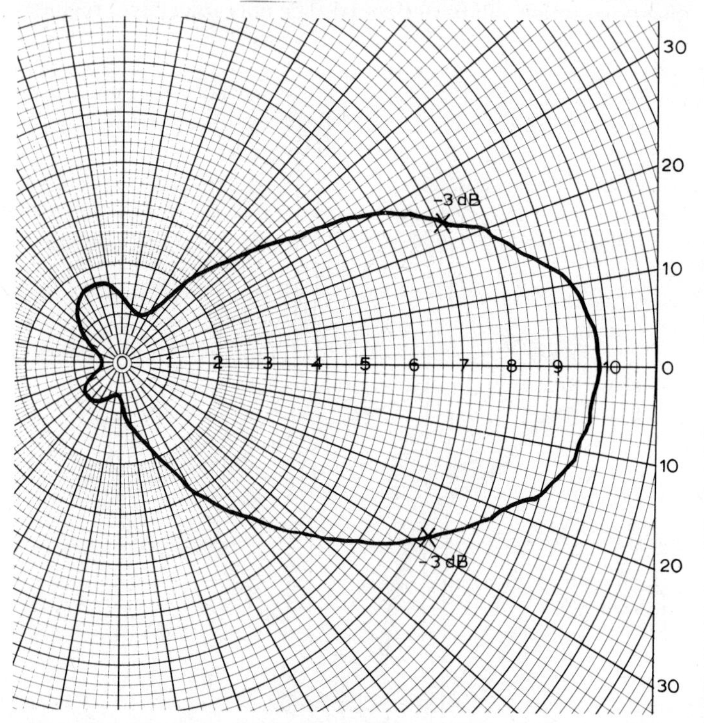

Fig. 3.32. Radiation pattern of the double diamond antenna in vertical mode. Forward gain is approximately 8 dB over a dipole

dipole is about 8 dB which is quite considerable for a small beam antenna but which might well be increased by the use of parasitic directors of the same configuration, i.e. diamond shaped but consisting of a single loop.

Satellite operation

The various aspects of operation into space satellite repeater stations is not within the scope of this book. Information concerned with frequencies of operation and other relevant data can be obtained from the AMSAT organisation. As far as antennas are concerned, a simple dipole or ground plane will enable reasonably good signals to be obtained when a satellite is passing close overhead, but for DX work at horizon distances a low angle beam antenna should be considered. Compact beams such as the ZL Special designs given in Chapter 3 are suitable.

Most VHF operators use some form of beam antenna anyway but also with a small amount of tilt in an upward direction, which typically should be about 15°. For very high angle satellite passes, a dipole or crossed dipoles mounted above a reflector will give better results. The AMSAT organisations in various countries will provide ample literature on request, which will contain information not only on actual satellite operation but also on suitable antennas and about wave polarisation as well.

Reading list

Subject	*References*
Directional (beam) antennas VHF	*ARRL Antenna Book*, American Radio Relay League, available from Radio Society of Great Britain, 35 Doughty Street, London, WC1N 2AE.
As above	*Radio Amateur VHF Manual*, ARRL as above and available from R.S.G.B.
As above	*VHF-UHF Manual*, G. R. Jessop, R.S.G.B. as above.
Design and construction data h.f. antenna arrays but useful information that could be applied to VHF	*Beam Antenna Handbook*, W.I. Orr, Radio Publications Inc. (USA), available from R.S.G.B. as above.

As above but devoted to cubical quad antennas for h.f. bands	*Cubical Quad Antennas*, W.I. Orr, Radio Publications Inc. (USA), available from R.S.G.B.
Log periodic antenna design	*Ham Radio Magazine*, (USA), May, 1975.
Antennas for satellite communication	*Ham Radio Magazine,* (USA), May, 1974.
Compact 2m beam antennas	*Practical Wireless*, (UK), May, 1977.
Loop Yagi antennas	*Ham Radio Magazine*, (USA), May, 1976.
Yagi antenna design	*Ham Radio Magazine*, (USA), August, 1977.
Portable quad antenna	*Radio Communications*, (R.S.G.B.), April, 1977
The VHF Quagi (Yagi quad hybrid system VHF and UHF)	*QST*, (USA), Journal of ARRL, April, 1977

Chapter 4

Matching and Feed Cables

Transmission lines

Virtually any cable that carries power from a source to a load can be regarded as a 'transmission' line. For example, a pair of wires connected to a battery at one end and a lamp at the other could be regarded as such. If the transmission line is long, power will be lost because of the self resistance of the line, and this applies equally when feeding power from a transmitter to an antenna. In this case however, we are also dealing with alternating current and not pure d.c., so other problems arise because, at radio frequencies especially, transmission lines have self inductance and capacitance which present reactance. With the combination of inductive and capacitive reactance plus resistance, all transmission lines therefore have what is known as a characteristic impedance. Since we are dealing with transference of power at different frequencies, other factors such as wavelength and velocity also come into the picture and these to some extent control the physical properties of transmission lines. Although purely resistive losses cannot be completely nullified – these are inherent in all conductors – steps can be taken to prevent radiation loss by suitable construction and by careful matching of the line to the source and the load. Loss due to radiation can be prevented by using two conductors arranged so that they are close enough together for the electro-magnetic fields from each to cancel out, i.e. so that they do not radiate, but this is possible only if the characteristic impedance of the line is equal to the impedance of both the source and the load.

The most common types of transmission line are open wire, which normally consists of two parallel conductors spaced a very small fraction of a wavelength apart, and coaxial cable, in

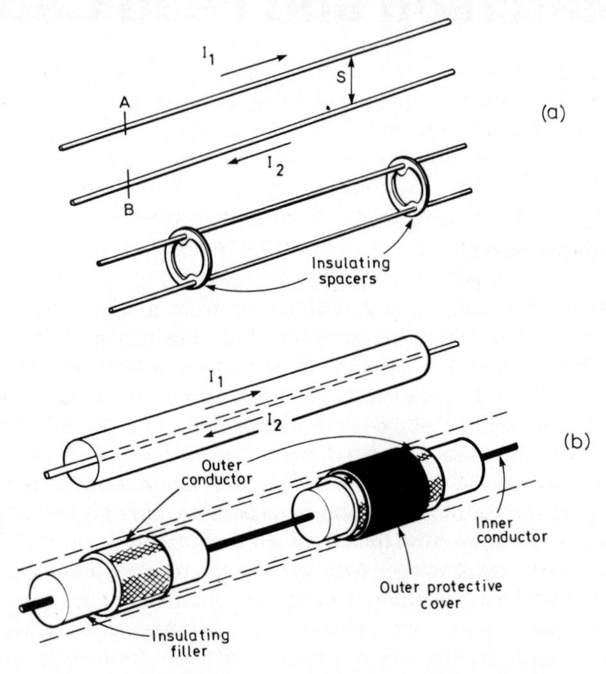

Fig. 4.1. (a) Open wire transmission line. (b) Coaxial transmission line (see text for explanation)

which one line is in effect shielded by the other, but otherwise behaves in much the same way electrically as an open two wire line. The configuration of each is shown in Fig. 4.1(a) in which currents I_1 and I_2 are flowing. If the current I_1 at point A on the open line has the same amplitude as current I_2 at the opposite point B, the fields set up by the two currents will be equal in amplitude and since they are flowing in opposite directions, the field from I_1 will be 180° out of phase with the field created by I_2. However, it takes time for the field from A to travel to B and vice versa, which means that the fields together may not be exactly 180° out of phase and so permit a small amount of radiation from the line. The best that can be done to prevent this is to have the two lines as physically close together as possible. In practice the space between the lines is made small consistent with the construction of the line and the frequency in use, although the higher the frequency the smaller must be the spacing. For instance, with a pair of open lines 150 mm (6 in) apart, the spacing is only a very small fraction of a wavelength at around 3 or 4 MHz, but at 144 MHz the phase difference introduced by the same spacing would be in the region of 26°, in which case the two fields would not cancel each other, at least not completely. A spacing of about 1% of the wavelength is tolerable with smaller spacing being desirable at VHF.

The second type of line, shown in Fig. 4.1(b) is the coaxial or concentric line consisting of one wire surrounded by its companion conductor in the form of a tube. The principle of operation is the same: because of close spacing the field due to current in the inner wire is cancelled by the field due to current in the outer conductor. Coaxial lines have another property, however, in that because of the *skin effect* (by which current flowing along the inner surface of the outer conductor does not penetrate through to the outer surface) the total field outside the cable is negligible because in effect the outer conductor behaves as a shield.

Line velocity

It is important to remember that although radio waves in free space travel at approximately 300 000 000 metres per second, they do not do so in conductors, i.e. the velocity is less. In coaxial lines the presence of dielectrics other than air reduce velocity very considerably. Open-wire air spaced conductors are not so affected and the velocity factor is generally near unity.

Since wavelength depends on velocity, the length of a transmission line will always be shorter in terms of wavelength than an equivalent length or distance in free space. A typical velocity factor for coaxial cable is 0.66 of the free space wavelength.

Characteristic impedance

Transmission lines have what may be regarded as distributed inductance and capacitance due to the fact that each unit length of line has its own small amount of capacitance and inductance, and one can think of a transmission line as shown in Fig. 4.2(a).

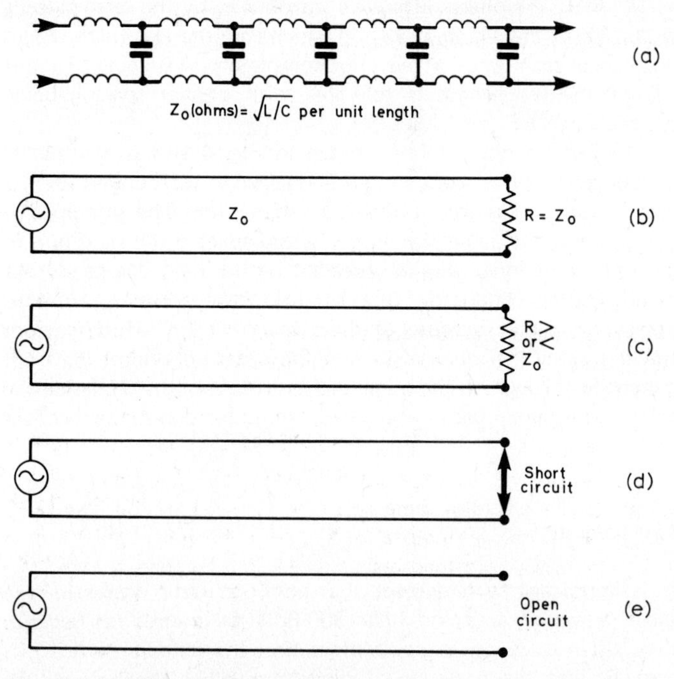

Fig. 4.2. (a) Distributed inductance and capacitance of a transmission line. (b)(c)(d)(e) Conditions that determine an impedance match or mismatch (see text)

Each small section of inductance limits the rate at which each small section of capacitance charges up and the effect of this is to establish a relationship between voltage and current. Thus the line has an apparent a.c. resistance or, as it is more commonly known, a characteristic impedance, the symbol for which is Z_0.

The inductive values decrease with increased conductor diameter and the capacitance values decrease with increased spacing between the lines. Hence a line with large diameter conductors closely spaced will have a relatively low characteristic impedance, whilst a line with wider spacing and thinner conductors will have a higher impedance. Impedances commonly used for open twin conductor lines are between 200 and 800 Ω, and for coaxial lines between 15 and 175 Ω, although the impedances most used in amateur radio applications are 50 and 75 Ω.

Line matching

If we take a transmission line of a given impedance and connect one end to a generator with the same impedance and the other to a purely resistive load R equal to the line impedance Z_0 as in Fig. 4.2(b), any current travelling down the line will flow into the resistance, which, as far as the current is concerned, is simply an extension of the line itself. The pure resistance has no inductive or capacitive reactance of its own and in this case the line is said to be perfectly matched as none of the power (I^2Z_0) generated by a travelling wave of current and voltage flowing down it and into the load is returned toward the generator. An infinitely long transmission line with the same impedance throughout its length would behave in the same way although power will eventually be dissipated by the natural resistance of the line. If we now take the case where the load resistance R does not equal that of the line impedance (Fig. 4.2(c)), then power which is not dissipated by the load is reflected back down the line. The greater the difference between R and Z_0 the smaller is the amount of power absorbed by R and so the mismatch becomes larger. Power flowing into R is called the *incident* power and power returning to the source is called *reflected* power. We can therefore talk of the ratio of reflected to incident power which becomes larger as the impedance mismatch increases. When R equals a short circuit as in Fig. 4.2(d) then all the power is returned as is also the case when R equals an open circuit as in (e).

When a mismatch occurs, power due to travelling waves exists in both directions along the line. The amplitude of the returned power is however dependent on phase differences between the incident and reflected voltages and currents which interact and set up what is known as a *standing wave* on the transmission line.

Standing waves

It has been illustrated that when a transmission line is correctly terminated there is no reflection of power and the flow of voltage and current is said to be a travelling wave. If the line is not correctly terminated the voltage to current ratio is not the same for both the load and the line, so power fed along the line will not all be absorbed by the load. Some will therefore be reflected as a returned travelling wave. These two, forward and reflected waves, interact along the line and as a consequence set up a standing wave. Fig. 4.3 illustrates this and in (a) an open circuit at the end of the line prevents current flowing out of it. The current waveform at this point has zero amplitude and thus in effect cancels itself because of a reversal of polarity – i.e. the phase is reversed. The current is travelling along the line but the voltage is across the line and so is not reversed by this reflection. The electric fields of the two waves, forward and reflected, therefore add to twice the amplitude. Fig. 4.3(b) and (d) denote conditions where R is greater or smaller than Z_0 and thus produce a standing wave of lower amplitude as in both cases only part of the forward power point is reflected. Fig. 4.3(c) shows the situation when $R = Z_0$, i.e. when no power is reflected since it all flows into the load and we have a uniform travelling wave. The remaining diagram (e) shows a short circuit across the line, and here the standing wave amplitude is the same as for the open circuit condition except that it has moved along to meet the conditions of zero voltage (V_{min}) at the short circuited end of the line.

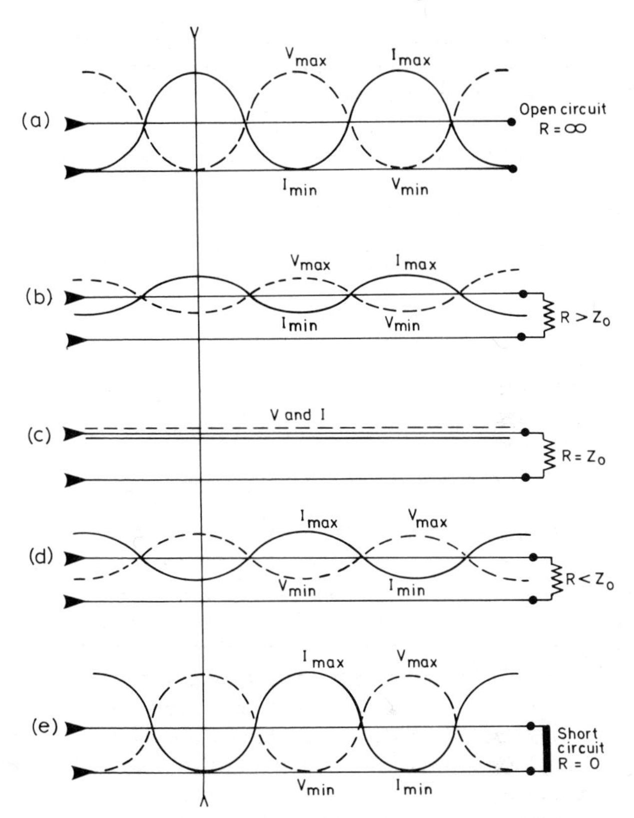

Fig. 4.3. Formation of standing waves when (a)(b)(d)(e) the impedance of the load, does not match that of the line; (c) A travelling wave when the load $R = Z_0$ of the line

The ratio of maximum (peak) voltage to minimum (trough) of the standing wave is called the *voltage standing wave ratio* (VSWR) and this is equal to the ratio of the mismatch R/Z_0 when R is greater than Z_0, or Z_0/R when R is less than Z_0. If the match is perfect, and this is rarely ever the case, then the VSWR equals 1 to 1. When a mismatch occurs the VSWR is greater and becomes more or less infinite with an open or short circuit condition, which incidentally should be avoided as high reflected power can damage transistorised transmitter output stages.

Types of transmission line

Whilst there are two basic types of transmission line, there are a number of variations of these – for example spaced twin wire lines can have virtually air spaced insulation between them, or they can be constructed with insulation of polythene, or similar material along the whole length. In the latter case, the line is usually of low characteristic impedance requiring the two wires to be very closely spaced. There are also three and four wire transmission lines but they have little or no application for the VHF operator, except perhaps for special types of antenna. The most popular transmission line in use today is the coaxial cable,

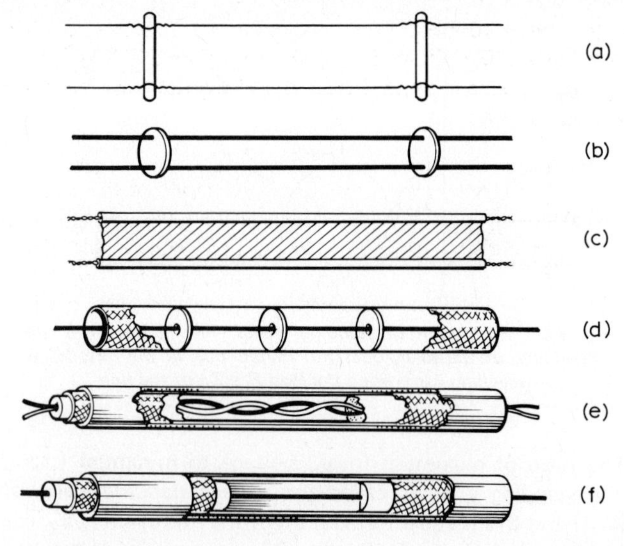

Fig. 4.4. Construction of various types of transmission line. (a)(b) Open wire; (c) 300 Ω ribbon feeder; (d) Air spaced coaxial; (e) Semi air-spaced coaxial; (f) Solid dielectric coaxial

which, although exhibiting higher loss than open wire lines, is, for the radio amateur, much more convenient to use as it requires no separate insulation from supports between transmitter and antenna; it is, moreover, flexible.

Spaced wire transmission lines (Fig. 4.4(b) and (b) are rarely used except for very large antenna arrays, generally on account of the difficulties involved in supporting and keeping the lines completely insulated during the run from transmitter to antenna. The air spaced line however, has the lowest loss per given length of all transmission lines but at VHF calls for fairly large diameter conductors very closely spaced even for impedances as low as 500 to 600 Ω. Some idea of the physical requirements can be obtained from Fig. 4.5 which gives the characteristic impedance for various wire diameters and spacing. This chart is based on the formula $276 \log_{10} D/d$, where D is the spacing between conductors (centre to centre) and d is the diameter of the wire. For example, with two conductors spaced 50 mm (2 in) apart and of diameter 6 mm (0.25 in), the characteristic impedance will be $276 \log_{10} 50/6.25 = 332\ \Omega$ (approx). From the chart in Fig. 4.5 and for the spacing/diameter ratio of 8, the graph shows the same impedance as at X.

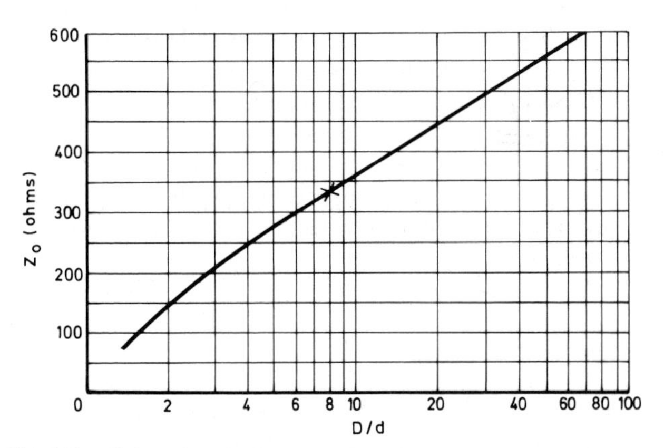

Fig. 4.5. Graph for determining the impedance of open line transmission lines with respect to spacing D and diameter of conductor d (based on 276 $\log_{10} D/d$)

The alternative to the air spaced twin wire transmission line is what is popularly known as 'ribbon feeder', which consists of two, usually stranded, wire conductors, with surround polythene insulation and a web of the same material all the way along to keep the wires equally spaced at about 12 mm (½ in) apart (Fig. 4.4(c)). The most commonly used version of this type of feeder has an impedance of 300 Ω, but there is a similar moulded insulation twin line feeder for 70 Ω. These cables are not much used for 2 metre operation since the majority of commercial transceivers (Japanese black boxes) have a coaxial output with an impedance of 50 Ω. With antennas having a different impedance, it is usual to employ a suitable matching device and thus retain the convenience of a 50 Ω coaxial feed cable.

Coaxial cables are available in a variety of forms, such as air spaced (very expensive (Fig. 4.4(d) and therefore little used by radio amateurs) and what might be called semi-air spaced (Fig. 4.4(e), as the stranded wire conductor is interwoven with a polythene line and loosely enclosed in a tube of similar material. This type of cable is more expensive than the better grade solid dielectric cables but is well worth considering for long cable runs where losses can be high (especially at 432 MHz). The remaining types of coaxial cable are the solid insulated variety (Fig. 4.4(f)) consisting of a stranded or solid inner conductor moulded within the dielectric. With the better quality types, losses are acceptable except for very long runs, for example, more than 20 m. It does not pay however to use low grade cables even for short runs, and some low grades have as much as 6 dB per 30 m loss at 145 MHz. This would mean that a beam antenna with a gain of 6 dB used with such cable would finally radiate little more power than a dipole with very low loss cable.

Matching devices

Inaccurate matching between the feed cable and the antenna results in high VSWR and therefore a loss of radiated power. The most common transmission lines in use for VHF are coaxial types of 50 or 70 Ω impedance and possibly 300 Ω ribbon feeder, which will operate directly with many types of antenna but, with some, may require special matching devices. A balanced feed is often necessary with coaxial cables if only to prevent r.f. returning along the outer screening braid. For example a 70 Ω coaxial cable will match closely into a dipole antenna, but a dipole is a balanced load and ideally requires a balanced feed. Single conductor coaxial lines are unbalanced. A simple method of achieving a balanced

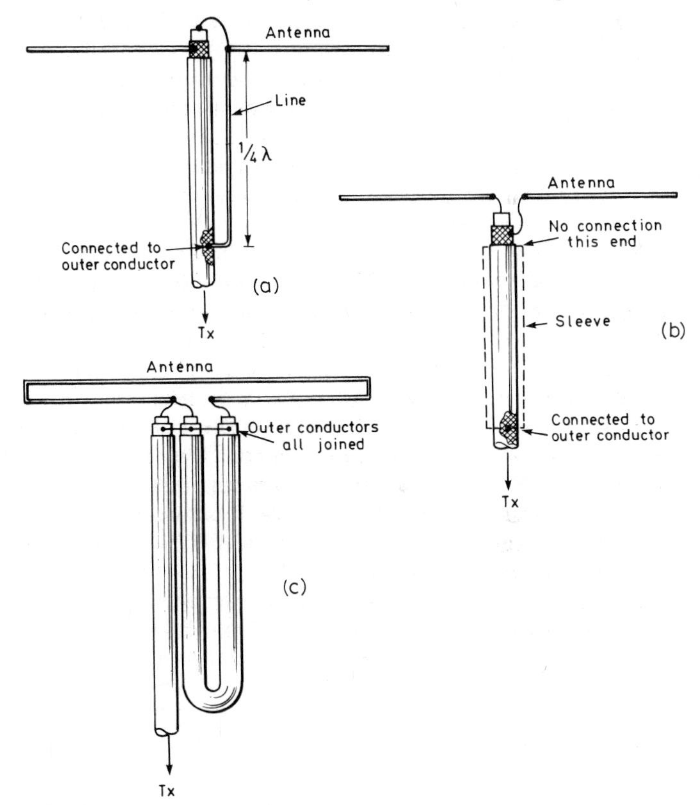

Fig. 4.6. (a) Quarter-wave line section to obtain a balanced feed to a dipole from unbalanced coaxial cable. (b) Quarter-wave sleeve to obtain a balanced match as (a). (c) Sometimes known as a 4 to 1 balun due to the impedance step up ratio it provides as well as a balanced feed

connection is to employ a balance to unbalance sleeve or line section sometimes referred to as a 'Bazooka' line balancer. It can take two forms as shown in Fig. 4.6 in which one as (a) is a $\frac{1}{4}\lambda$ section of line in parallel with the last $\frac{1}{4}\lambda$ portion of the cable, or (b) as a $\frac{1}{4}\lambda$ sleeve also fitted over the last $\frac{1}{4}\lambda$ portion. The method is to present a high impedance to r.f. at the top of the cable and prevent any current flow down the outer conductor of the line. The length of line or sleeve is a full quarter-wave so velocity factor plays no part in this application. Note that with the sleeve type there is no connection between the sleeve and the coaxial cable at the top, i.e. near the antenna. Only the bottom end of the sleeve is connected to the braiding of the coaxial outer conductor. The sleeve may in fact be mesh braiding from a larger diameter piece of coaxial cable slid over the insulating outer covering of the antenna feed cable.

Another form of balance to unbalance feed device commonly called the 'balun' (a coined name from 'balance to unbalance'), consists of a half-wave section of feed cable of the same impedance as the main feed cable coupled as shown in Fig. 4.6(c). This provides an impedance step up of about 4 to 1 from its self impedance. In this case however, the half-wave length of line must take the velocity factor of the cable into account and which for nearly all coaxial cable is 0.66. The length of cable required for the loop (see equations on page 17) will therefore be $(150 \times 0.66)/f$ metres. A balun of this nature could be used to feed a folded dipole directly from a 50/70 Ω coaxial cable with a reasonably good match.

There is another form of balun which can be made to provide impedance step up. This is a capacitive/inductive device but is rather difficult to make and adjust. A circuit for use at 145 MHz is shown in Fig. 4.7(a) together with details of component values. This will match balanced loads in the region of 100 to 1600 Ω to a coaxial line of 50/70 Ω. Some adjustment of the tapping points on L_2 may be necessary for impedances at the output of lower than 100 Ω, i.e. they will need to be moved nearer to the centre.

Note that the coil L_1 is wound centrally over L_2 and must be air-spaced as close as possible to L_2 without touching. For operation at 2 metres L_2 is five turns of 12 swg (2.64 mm dia.) tinned copper wire on a 12 mm ($^1/_2$ in) diameter former and pulled out to 21 mm ($^7/_8$ in) long. The taps are $1^1/_2$ turns in from each end. The coupling coil L_1 is two turns of 14 swg (2.03 mm dia.) tinned copper wire wound on a 25 mm (1 in) diameter former with 3 mm ($^1/_8$ in) spacing between turns. It can be placed centrally over L_2 with the required amount of air space around (Fig. 4.7(b)). The

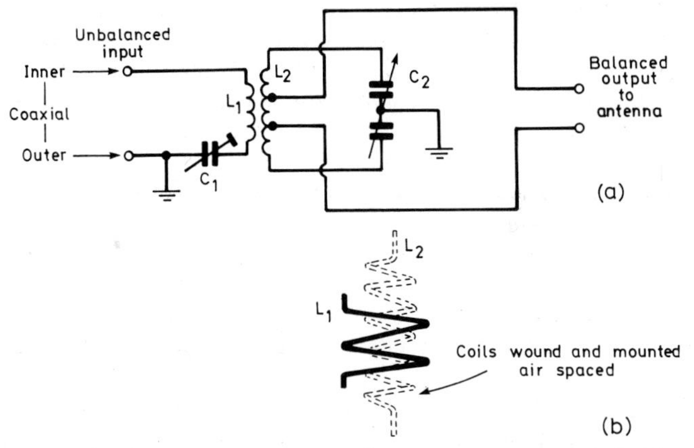

Fig. 4.7. (a) Circuit for inductive/capacitive balun for 145 MHz. (b) Note how the coil is wound centrally over L2 but with air spacing as close as possible to L2

complete circuit may be housed in a small aluminium box about 150 x 75 x 75 mm (6 x 3 x 3 in) with suitable coaxial sockets mounted at each end of the box.

The device must be used in conjunction with a VSWR indicator connected between the transmitter and the coaxial feed line. Adjust C_1 and C_2 to obtain the lowest possible VSWR reading; if this will not come down to a reasonable figure then the tapping points on L_2 may need to be changed to another position.

Common matching methods

There are many different ways of obtaining a match between a transmission line and an antenna. Those described here are among the most common likely to be used, although some have already appeared in the previous chapters as a more or less integral part of particular antennas (as for example the quarter-wave stub and the gamma match).

The delta match

With this method the transmission line, usually an open wire type, is opened out at the end where it meets the antenna, as shown in Fig. 4.8(a) and connected at equal distance points from

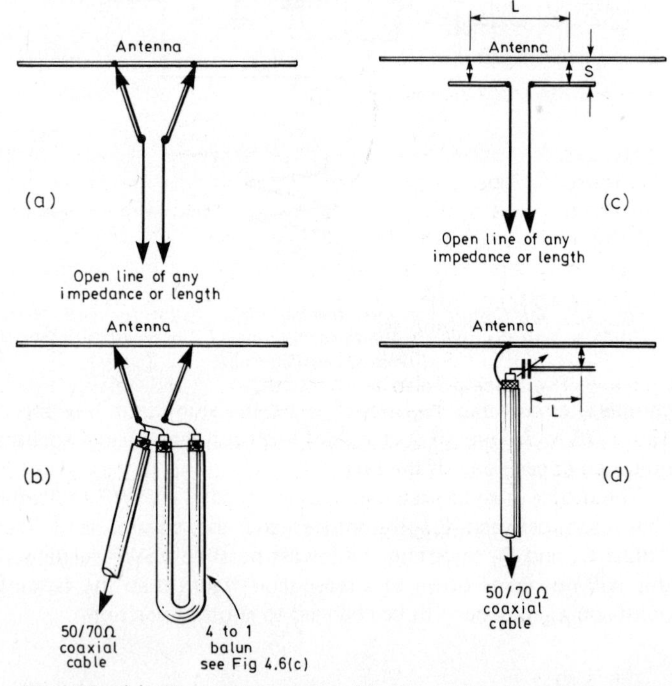

Fig. 4.8. (a) The delta match with open wire transmission line. (b) The delta match with coaxial transmission line and 4 to 1 balun. (c) The gamma match with open wire transmission line (see text for adjustment). (d) The gamma match with coaxial cable (see text for adjustment)

the centre. Although it has a slight disadvantage in that the opened out portion may radiate and it is critical as far as frequency is concerned, it is relatively easy to adjust and adaptable to open line or coaxial transmission lines. Adjustment for lowest VSWR is made by altering the length of the opened out sections and the point at which these tap into the antenna. When used with coaxial cables of 50/70 Ω impedance, a balun (4 to 1) can be used as in Fig. 4.8(b). Adjustment for minimum VSWR is the same and in either case the length of the 'delta' sides and the spacing between the tapping points will be in the region of 0.15 λ based on one wavelength in air.

The gamma and T-match

This is really a variation of the delta match with the true T-match being used for open wire lines as in Fig. 4.8(c) and the gamma match for coaxial lines as (b). The T-match method offers considerable flexibility in obtaining different impedance ratios and is also a balanced system. It could also be used with a 4 to 1 balun and 50/70 Ω coaxial cable. The input impedance increases as the length of the matching section L is increased, although beyond a certain point the input impedance decreases as L is made greater. The spacing S also has some bearing on the feed impedance. The section of the T-match L should be about 450 mm (18 in) long to allow adjustment of the variable position connections to the antenna element to obtain minimum VSWR. The spacing S will be about 50 mm (2 in) if thin tube is used for both the antenna and the T-match section.

The gamma match

This is shown in Fig. 4.8(d). It is very suitable for matching 50 to 70 Ω coaxial cable to the driving element of a parasitic array. The outer conductor of the coaxial cable is connected to the centre point of the antenna, and the inner via a small capacitor to a short matching section – the length of which will be about

100 mm (4 in) for 145 MHz. This section could be made a little longer and a variable tap provided in the same way as for the T-match. The capacitor, which will be air-spaced and about 25 pF for 145 MHz, is used to tune out the reactance of the matching section. The connection between the matching section and the antenna and the capacitor are adjusted to obtain minimum VSWR.

Other methods of matching

Some methods used are effectively part of the antenna design but often derived from one of the basic principles in the preceding paragraphs. There is also the 'Q' match which is a section of transmission line of a given impedance often used as a 'transformer' to obtain a high to low impedance match. It is not used much in the more popular types of antennas for VHF, but details on its operation and construction can be found in the references given at the end of this chapter.

The folded dipole as a matching device

The folded dipole can be used as a matching device even though it may actually operate as a driven element in a parasitic array, as is often the case. A (2-element) folded dipole as in Fig. 4.9(a) has a centre impedance of about 300 Ω, but in the presence of other conductors, as in a parasitic array, this impedance may be reduced to as low as 70 Ω, so that a direct match could be made to a coaxial transmission line of the same impedance. If a feed impedance of higher than 300 Ω is required, for example to match a 600 Ω open line with a folded dipole used by itself, the impedance at the centre can be raised to about 630 Ω by the addition of another parallel element, as in Fig. 4.9(b), and this is close enough to 600 Ω for a good match. Additional parallel elements will raise the impedance even higher, but a better method of achieving this is to use two elements with one of larger diameter than the other which, in conjunction with the spacing between them, can be

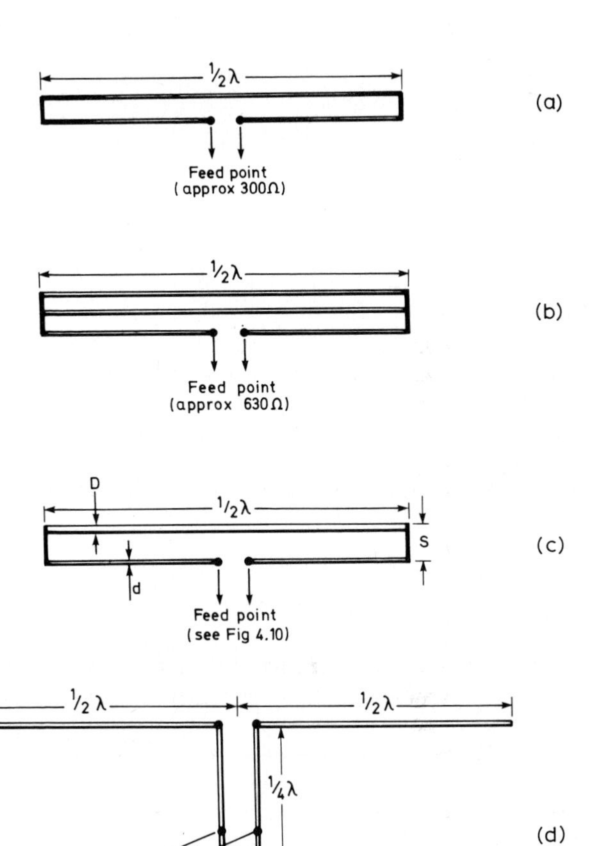

Fig. 4.9. (a) The folded dipole permits a direct feed of 300 Ω with ribbon feeder or coaxial cable used with a 4 to 1 balun. (b) A three wire folded dipole permits a direct feed from a 600 Ω open transmission line. (c) A folded dipole with element diameters and spacing variation permits a range of feed impedances as outlined in the text and in Fig. 4.10. See text with reference to (d)

utilised to provide impedance step up ratios of between about 2 and 15 to 1. The configuration is shown in Fig. 4.9(c) and the impedance ratios that can be obtained with element diameter differences and spacing can be found from Fig. 4.10. Two known factors are required to obtain the impedance ratio. These are *D/d*, where *D* is the diameter of the thicker element and *d* that of the

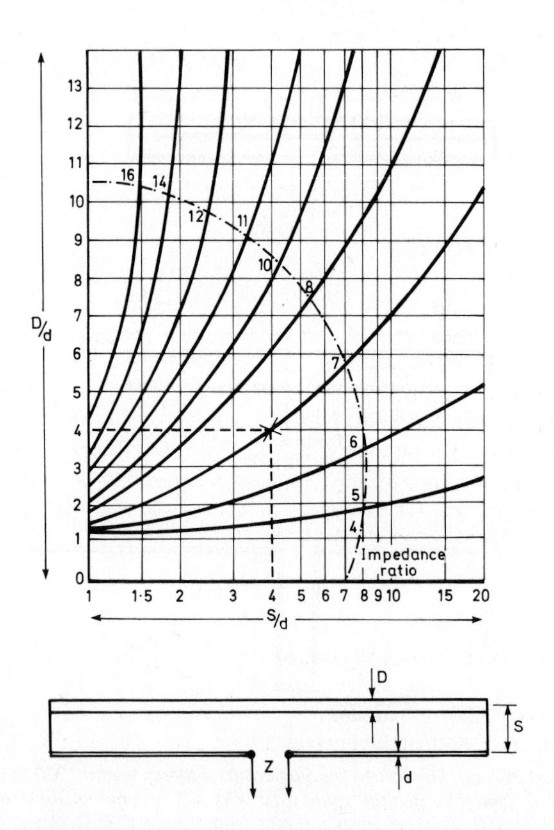

Fig. 4.10. Different impedance ratios can be obtained by variation of element diameters and spacing

thinner element. The spacing S between the two is divided by D to obtain the other factor, i.e. S/D. Using the graphs of Fig. 4.10, the example given by the dotted lines is taken from

$$\frac{D}{d} = \frac{24\ \text{mm}}{6\ \text{mm}} = 4 \quad \text{and} \quad \frac{S}{d} = \frac{100\ \text{mm}}{25\ \text{mm}} = 4$$

which is an impedance step up of 7 to approximately 2100 Ω. However, let's suppose that with a folded dipole in the presence of other conductors, as in a parasitic array, its natural impedance is reduced to, say, 40 Ω. The use of uneven element diameters and spacing as above would raise the impedance to 40 × 7 or 280 Ω and therefore provide a good match to 300 Ω ribbon feeder.

The quarter-wave stub

Open ended and closed end $^1/_4\lambda$ stubs are frequently used as matching devices, and the most common application is shown in Fig. 4.9(d) in which two $^1/_2\lambda$ dipoles are to be fed in phase as a collinear antenna. The stub provides a means of obtaining a low impedance feed (50 to 70 Ω) to the otherwise very high impedance (due to end feeding each dipole). The feed points are adjusted experimentally, depending on the impedance of the cable used, to obtain the lowest VSWR. The $^1/_4\lambda$ stub is used in many of the antennas described in Chapters 2 and 3.

Coaxial cables, plugs and connections

As coaxial cables are used more than any other type of transmission line for VHF and UHF operation, this book would not be complete without details of the most popular coaxials used and also a few words on cable connections. The following lists provide relevant data concerned with characteristic impedance, BICC and Service Numbers and most important the loss in dB per 100 ft/MHz (and metric equivalent).

COAXIAL CABLES DATA

Impedance group		50	50	50	50	50	50	50
BICC list number		TT3346 TT3345	—	—	—	—	—	T3234
Service number BS2316-UNIRADIO (UR)		74/75	4	67	43	76	115	—
Characteristic impedance (Ω)		51	46	50	52	51	52	52
Nominal capacitance	(pF/ft) (pF/m)	30.7 100	33 108	30 98	29 95	29 95	29 95	22.5 74
Nominal velocity ratio		0.664	0.666	0.666	0.674	0.674	0.674	0.82
Nominal attenuation at (dB/100 ft/MHz) (dB/30m/MHz)	MHz 100 200 300 600	 0.96 1.45 1.86 2.93	 2.26 3.45 4.33 7.48	 2.07 3.02 3.8 5.66	 3.95 5.69 7.06 10.3	 5.1 8.04 10.9 19.3	 3.95 5.69 7.06 10.3	 2.6 3.6 4.5 6.5
Maximum power rating at (kW/in air/MHz)	MHZ 100 200 300 600	 1.9 1.3 0.99 0.64	 0.6 0.42 0.3 0.23	 0.54 0.37 0.3 0.2	 0.13 0.185 0.23 0.34	 0.13 0.082 0.06 0.034	 0.17 0.12 0.098 0.068	 0.20 0.14 0.11 0.08
Maximum r.f. voltage (kV peak)		15	4.75	4.8	2.75	1.8	2.75	0.7
Cable outisde diameter	(inches) (mm)	0.870 22.1	0.405 10.3	0.405 10.3	0.195 5.0	0.195 5.0	0.285 7.2	0.270 6.9
Suitable amateur radio VHF/UHF		VHF and UHF	VHF short runs	VHF	VHF short runs	VHF short runs	VHF short runs	VHF

COAXIAL CABLES DATA (Contd.)

Impedance group		75	75	75	75	75	75	75
BICC list number		—	T3365	—	—	—	T3141	—
Service number BS2316-UNIRADIO (UR)		77	—	!	57	59/65	—	39
Characteristic impedance (Ω)		75	72	72	75	75	75	69
Nominal capacitance	(pF/ft)	20.5	21.3	21.2	20.6	20.6	20.5	23
	(pF/m)	67	70	69	68	68	67	75
Nominal velocity ratio		0.664	0.666	0.666	0.666	0.666	0.666	0.674
	MHz							
Nominal attenuation at	100	0.95	1.7	1.74	1.87	1.87	3.1	2.5
(dB/100 ft/MHz)	200	1.45	2.45	2.65	2.74	2.74	4.4	3.63
(dB/30m/MHz)	300	1.86	3.1	3.34	3.45	3.45	5.5	4.54
	600	2.93	4.5	5.67	5.17	5.17	8.1	6.72
Maximum power	MHZ							
rating at	100	1.6	0.65	0.64	0.55	0.55	0.31	0.36
	200	1.05	0.45	0.42	0.375	0.375	0.22	0.25
(Kw/in air/MHz)	300	0.84	0.35	0.33	0.3	0.3	0.17	0.2
	600	0.545	0.24	0.2	0.2	0.2	0.12	0.13
Maximum r.f. voltage (kV peak)		12.5	6.5	6.25	5.0	5.0	3.0	4.0
Cable outisde diameter	(inches)	0.870	0.642	0.450	0.405	0.405	0.310	0.310
	(mm)	22.1	16.3	10.3	10.3	10.3	7.9	7.9
Suitable amateur radio VHF/UHF		VHF and UHF	VHF and UHF short runs	VHF and UHF short runs	VHF and UHF short runs	VHF and UHF short runs	VHF	VHF

COAXIAL CABLES DATA (Contd.)

Impedance group		50	50	50	50	75	75	75
BICC list number		—	—	—	—	—	—	—
Service number RG/U Type		RG58U	RG8U	RG17U	Foamed RG8AU	RG59U	RG11U	RG11
Characteristic impedance (Ω)		53.3	52	52	50	73	75	75
Nominal capacitance	(pF/ft)	—	29.5	—	—	—	—	20.5
	(pF/m)	—	97	—	—	—	—	67
Nominal velocity ratio		0.659	0.659	0.659	0.75	0.659	0.659	0.666
	MHz							
Nominal attenuation at	144	6	2.5	1	2	4.2	2.8	Similar
(dB/100 ft/MHz)	220	7	3.5	1.3	2.75	5	3.7	←
(dB/30m/MHz)	420	15	5	2.3	3.9	8	5	RG11U
Maximum power	MHZ	Watts	Watts	Watts	Watts	Watts	Watts	
rating at	144	175	800	2300	800	250	800	—
(Watts/in air/MHz)	220	135	650	1900	650	180	650	—
	420	90	400	1200	400	125	400	340
Maximum r.f. voltage (kV peak)		—	—	—	—	—	—	5
Cable outisde diameter	(inches)	0.195	0.405	0.87	0.405	0.242	0.405	0.405
	(mm)	5.0	10.3	22.1	10.3	6.1	10.3	10.3
Suitable amateur radio VHF/UHF		VHF short runs	VHF and UHF short runs	VHF and UHF	VHF	VHF short runs	VHF	VHF

CABLES TO UNITED STATES MILITARY SPECIFICATION MIL-C-17D WITH CORRESPONDING BICC LIST NUMBERS AND BS 2316 APPROXIMATE EQUIVALENTS

MIL-C-17D reference	*BICC List No.*	*BS 2316 approximate equivalent*		*Impedance (Ω)*
		UR No.	*URM No.*	
RG 6A/U	T3351	–	–	75
RG 11A/U	T3373	UR57	URM57	75
RG 34B/U	T3339	–	–	75
RG 35B/U	T3340	–	–	75
RG 58C/U	T3352	UR76	URM76	50
RG 59B/U	T3353	UR90	URM90	75
RG 113A/U	T3393	UR78	–	95
RG 164/U	T3332	UR77	URM77	75
RG 177/U	T3343	–	–	50
RG 212/U	T3297	–	–	50
RG 213/U	T3382	UR67	URM67	50
RG 214/U	T3298	UR112	URM112	50
RG 215/U	T3381	–	–	50
RG 216/U	T3396	UR60	URM60	75
RG 218/U	T3345	UR74	URM74	50
RG 219/U	T3346	UR75	–	50
RG 220/U	T3347	–	–	50
RG 221/U	T3348	–	–	50
RG 223/U	T3354	UR115	URM115	50
RG 224/U	T3349	–	–	50

By courtesy of BICC Ltd.

Coaxial plugs and connections

The most widely used is the PL295 of which there is one type for large diameter cables and one for small, e.g. cable of not more than about 6 mm (¼ in) diameter. It is always very important that these and other types of VHF/UHF plugs are properly connected otherwise losses can occur. The correct way of connecting PL295 plugs with or without cable diameter adjusters is shown in Figs. 4.11 and 4.12. The captions explain the procedure.

Where the far end of a coaxial cable joins the antenna it is vital that protection is provided to prevent moisture reaching the connections and also the end of the cable. If moisture creeps into a coaxial cable it can cause high losses and indeed completely ruin the performance of the cable either due to corrosion of the inner

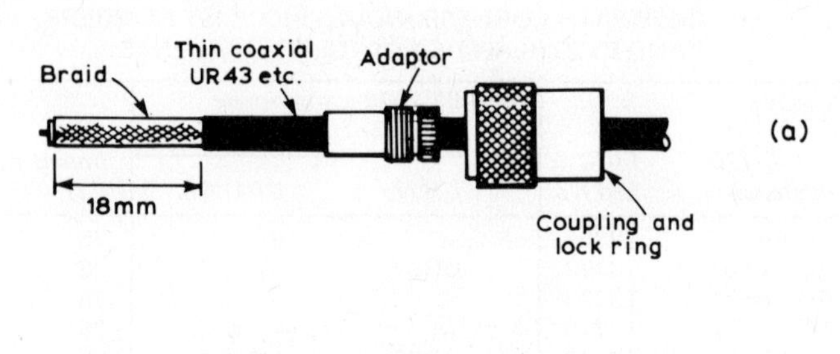

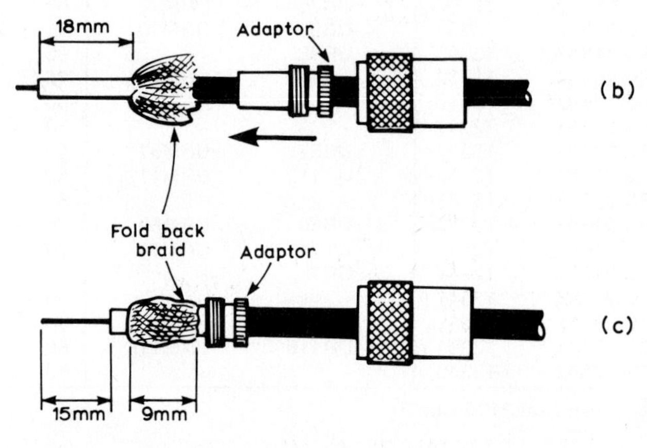

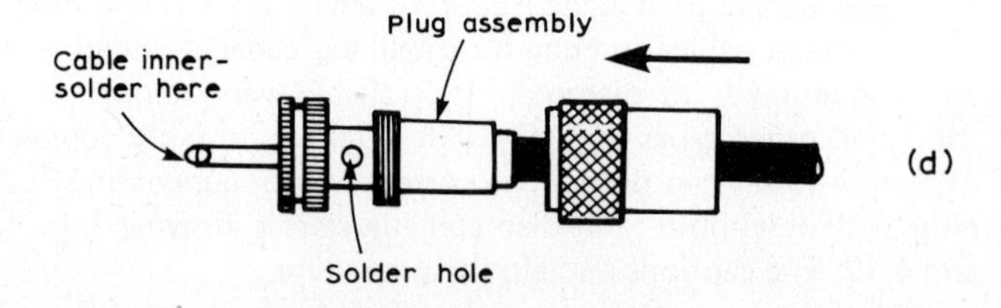

Fig. 4.11. (a) Cable trimmed and adaptor and coupling ring fitted. (b) Braiding folded back. (c) Adaptor brought down inside folded braid. (d) Plug assembly fitted neatly for soldering as shown

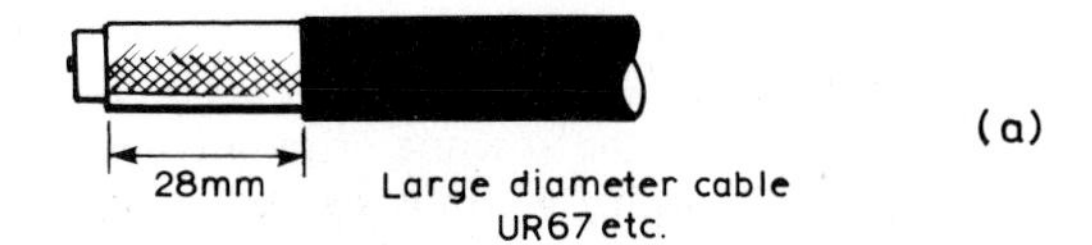

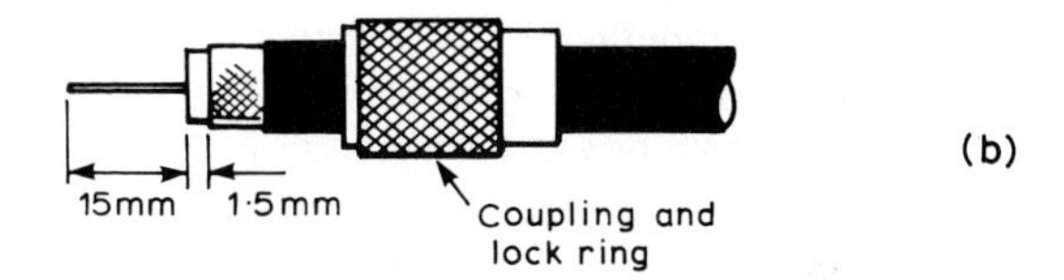

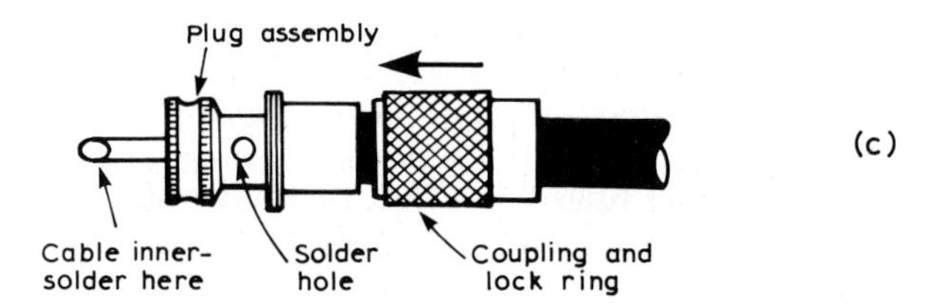

Fig. 4.12. (a) Cable trimmed ready for (b). (b) Inner conductor prepared and lock ring fitted. (c) Plug assembly fitted and soldered as shown

and outer conductors or simply by its presence within the cable. Silicon grease over the exposed end of the cable is also a worthwhile extra precaution against the entry of moisture. Sharp bends should be avoided when installing coaxial cable and care taken in handling to avoid kinking the cable which can fracture or even break a solid wire inner conductor. Do not use metal cleats to secure cables to walls etc., but at the same time a cable should not be freely suspended so that its whole weight is held by its connections or a single clamp at the antenna. Thin nylon ties or plastics clips should be used every few feet to provide support along the full length. It is also prudent not to cut a cable that is a metre or so too long for the run. There is always the possibility that one may later require to move the antenna or raise its height, and in the meantime a small surplus will not add much to total cable loss. When there is more than one antenna at the mast top, as is frequently the case, the cables from each should not be run close together, particularly when the antennas are for the same frequency. For example it is common practice to have a beam antenna and an omnidirectional antenna together. Interaction between cables from each can give rise to distortion of the radiation pattern from one or the other, or both.

Antenna rotators

This is an item that is best purchased, but it must be chosen with due regard to the weight it will have to carry. Some radio amateurs build their own rotators, but to do so does entail a good deal of precision mechanical work. If one or more beam antennas are to be carried, e.g. for h.f. bands and VHF/UHF bands, then a sizeable rotator such as that shown in Fig. 4.13 will be required. There are smaller versions capable of carrying small beams for VHF or UHF, but it is always worth considering what future requirements might be. Commercially made rotators are normally supplied with a control box that permits stopping at a pre-selected point and/or continuous rotation through 360° with automatic stop at the zero/360° mark, according to the direction of rotation. Some rotators (like that shown in the photograph) are arranged so that the under section is clamped to the top of the mast as in Fig. 4.14(a) with the upper section carrying a stub mast long enough to mount one or two beam antennas of relatively light weight and small dimension, such as VHF/UHF arrays. The stub mast section should be of insulating material, particularly if the beam antenna is operated vertically. A metal stub mast can seriously distort the radiation pattern as well as reduce the efficiency of the antenna. If large beams are to be carried then a better method could be used, as shown in Fig. 4.14(b), to take the strain from the rotator.

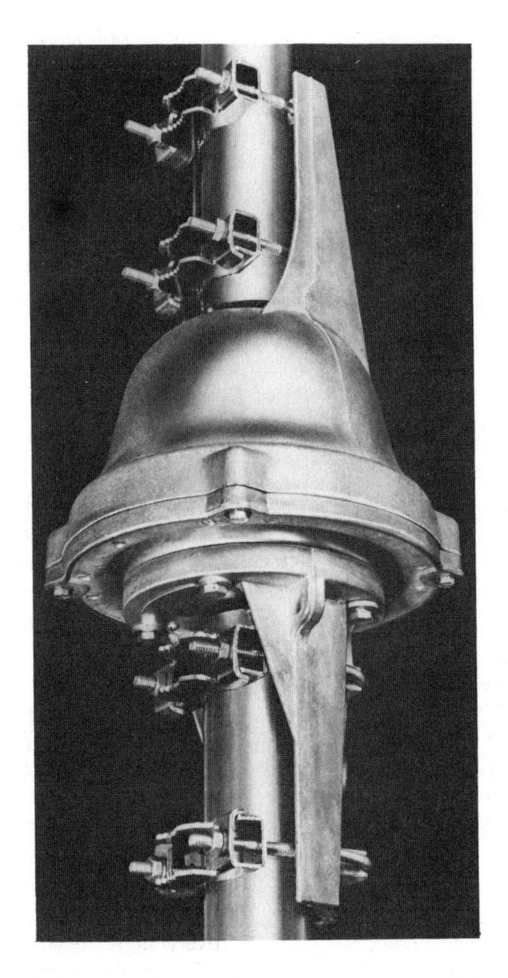

Fig. 4.13. Typical high power antenna rotator (photo by courtesy of Aerialite Limited)

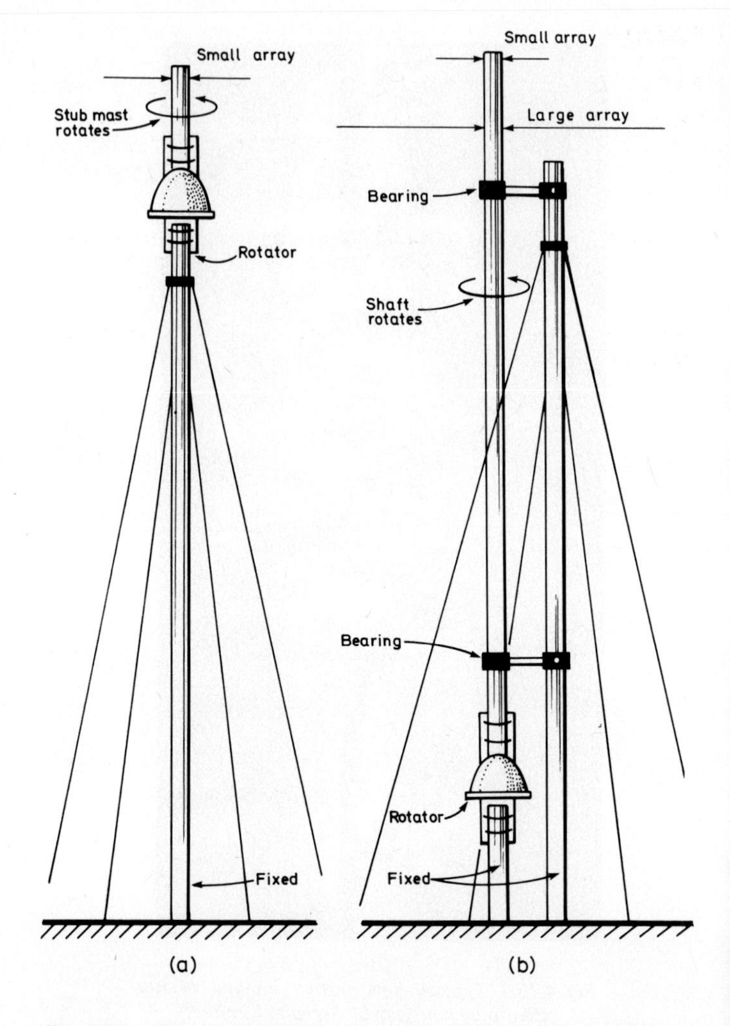

Fig. 4.14. The methods of using a rotator. (a) At top of mast with short stub and suitable for small arrays. (b) At bottom of long shaft with bearings on main mast when large and small beams are carried together

Reading list

Subject	*References*
Matching and transmission lines	*ARRL Antenna Book*, American Radio Relay League, available from the Radio Society of Great Britain, 35 Doughty Street, London WC1N 2AE.
As above	*Radio, TV and Audio Technical Reference Book*, S. W. Amos, Newnes-Butterworths, 1977.
Feeding and matching techniques VHF/UHF	*Ham Radio Magazine*, (USA), May, 1976.

Chapter 5

Antenna Performance

Measuring the performance of an antenna is a difficult task even with sophisticated equipment because of the number of variable factors involved, most of which cannot always be reasonably accounted for. Measurement of antenna gain for example is only possible with any degree of accuracy when made under ideal environmental conditions – and with virtually perfect matching between both the antenna being tested and its transmission line and also the reference antenna and the same transmission line, which must be used so that line loss is the same in each case. Also when one antenna is substituted for another, the position must be the same. The ideal environment is of course free space, so that the wave travelling between the antenna being tested and/or the reference antenna is not influenced in any way by reflection from ground or surrounding conducting objects. With antennas operating in the h.f. region, i.e. below about 50 MHz, direct performance measurement of any kind, except possibly of field strength over the surrounding terrain, is nearly impossible, and so antenna designers frequently resort to use of VHF or UHF models so that 'free space' conditions can be more closely approximated.

All antennas behave in exactly the same way regardless of the frequency at which they are operated, for example, a $\frac{1}{2}\lambda$ dipole antenna cut to operate at say 7 MHz would have the same figure-of-eight radiation pattern as a dipole cut to operate at 70 or

700 MHz. This scaling down idea is used by designers of ships and aircraft who scale down the full size version to check performance of ships in wave tanks and aircraft in wind tunnels. Antennas also have the 'reciprocal' property in that the performance for receiving is the same as that for transmitting so tests can be made for gain and directivity when using an antenna in receiving mode. The 'free space' environment is much less difficult to simulate when using very high frequency scale models of antennas as the number of 'wavelengths' between ground and other objects becomes greater and therefore we need less space in dimensions of wavelength in which to work. Provided that a VHF or UHF antenna is sited at a fairly large number of wavelengths above ground it can be considered as being in free space. The distance between the measuring equipment antenna and the antenna being tested must also be a large number of wavelengths if reasonably accurate polar patterns are to be obtained. It is usual by the way to operate the antenna being tested in receiving mode so this will be connected to the measuring equipment. However, before one can even begin to think about gain and polar pattern measurements, it is essential to be able to check that the full amount of power is reaching the antenna from the transmitter, and that equally the full level of received signal is being transferred from antenna to receiver.

Cable loss and VSWR

The first and foremost requirement in maintaining antenna efficiency is to ensure that the VSWR is as low as can possibly be obtained, which is a reasonably sure sign that the transmitter output, the feed cable and antenna are matched with each other. There are potential problems other than mismatch between feed cable and antenna, and one may be a mismatch between transmitter output and cable. This is always a point worth checking with the aid of a power meter and dummy load which should of course have the same 'resistance' as the transmitter output impedance. Any cable used to connect one to the other must of course have the same impedance. Coaxial cable of unknown or doubtful quality, or cable into which moisture has seeped, can also result in a mismatch and/or high loss. It may therefore be false economy to use cheap cable, and where there is doubt some form of check should be made on the length to be used by connecting one end to the transmitter and the other to a dummy load. With a power meter in series with the load at the far end of the cable one can establish whether or not a power level reasonably close to that from the transmitter is being developed in the load. There should be only the few decibels loss introduced by the normal loss of the cable. External r.f. power amplifiers in circuit, but not in use, can introduce a higher than expected VSWR, as can coaxial switches, and these should be taken out of circuit when checking VSWR for the first time with a new antenna.

Acceptable VSWR readings are always a debatable subject and very low readings (e.g. close to 1 to 1 right across the band) should be regarded with suspicion, particularly when the feed cable, be this coaxial, is very long. The self loss of the cable which may be high (depending on its quality) will absorb power reflected from a mismatched antenna with a resultant low VSWR reading at the transmitter end. Ideally the VSWR should be checked at both ends of the cable particularly when trying to establish the goodness of match with the antenna.

Use of VSWR and power meters

Really accurate VSWR and r.f. power meters suitable for use at VHF are very expensive. Meters incorporated in transmitters should not be accepted as being accurate, and neither should the small but otherwise popular and relatively inexpensive Japanese made instruments for VSWR and/or power. If possible such instruments should be checked with another of known accuracy, although a rough idea of this can be obtained with a well made dummy load similar to that illustrated in Fig. 5.1(a), and connected as shown in (b). The dummy load consists of 10 or 12 resistors of such value that, when connected in parallel, as shown, will provide the requisite value of the load within ± 1 Ω. The resistors should be of 2 W rating for total power of not more than 20 W and must be of carbon construction to keep inductive

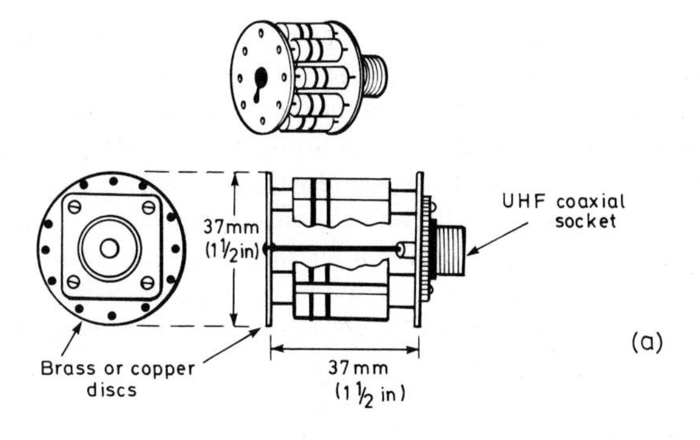

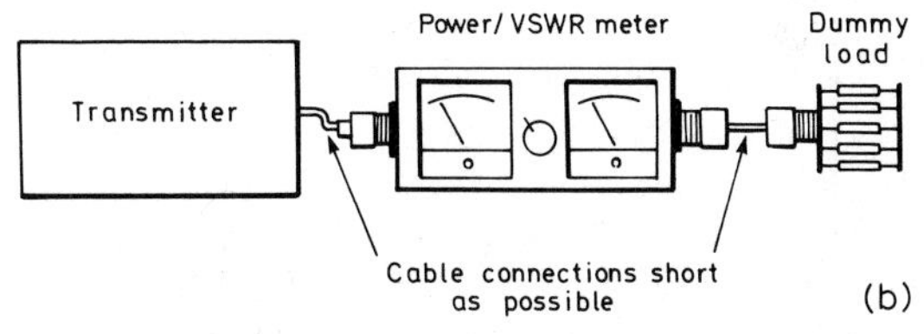

Fig. 5.1. Construction of a 50 Ω dummy load – see text with reference to resistor values

effects to an absolute minimum. Connections between VSWR meter and load should be as short as possible. If the load is made as illustrated the VSWR should be not greater than 1 to 1.1. It should, as with any test equipment, be checked if possible with a power meter of known accuracy.

Checking a new antenna

A newly made antenna should be checked with a short cable run at first to obviate possible false indications that might arise with a very long feed cable. A preliminary test on receive is worth while to ensure that the antenna is at least operating before applying power, and even then an initial check at low power is worth while if only to prevent possible damage to transistor output stages in the transmitter, just in case of short or open circuit connections. A VSWR or power meter should be connected in the line between transmitter and feed cable if such a meter is not already incorporated in the transmitter. If forward power at the right level or a reasonably low VSWR is indicated, then higher power can be applied to make the final check, although any tuning required at the antenna should be done at lower power. External high power amplifiers should never be connected in circuit and operated with an antenna until a low VSWR reading has first been established. If the antenna is fairly closely resonant at band centre then a check across the 2 metre band should reveal VSWR readings v frequency approximately as those in Fig. 5.2.

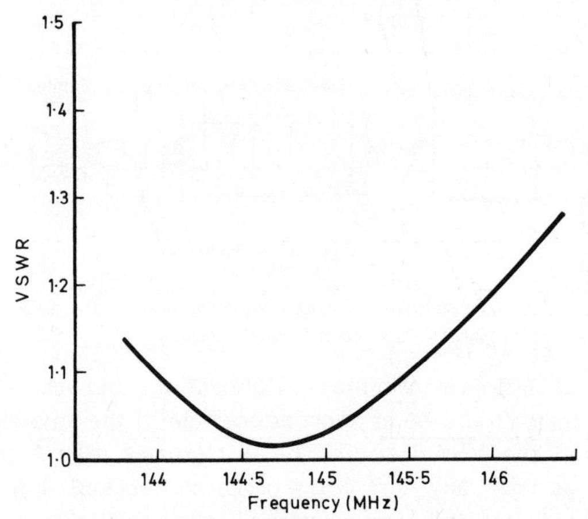

Fig. 5.2. Typical VSWR response curve with very low loss feed cable

They will nearly always be a little higher than shown as the 'perfect' match is rarely possible. However, there should be the slight rise at each side from band centre. Much will depend on cable loss however, for as already mentioned a very long cable or somewhat lossy cable even of short length may result in a near flat curve and almost certainly very low VSWR readings.

It is not generally known that VSWR readings obtained even with an accurate meter may not be true anyway and the true VSWR will generally be higher than the reading actually indicated.

This is due to cable loss, and often a cable with a high loss factor will produce a low VSWR reading which of course looks good even though there may be a large mismatch at the antenna end. With high cable loss the reflected power due to any mismatch at the antenna will be more greatly absorbed during its return journey down the cable end consequently the VSWR may appear to be low. The graphs in Fig. 5.3 give some idea of the corrections to

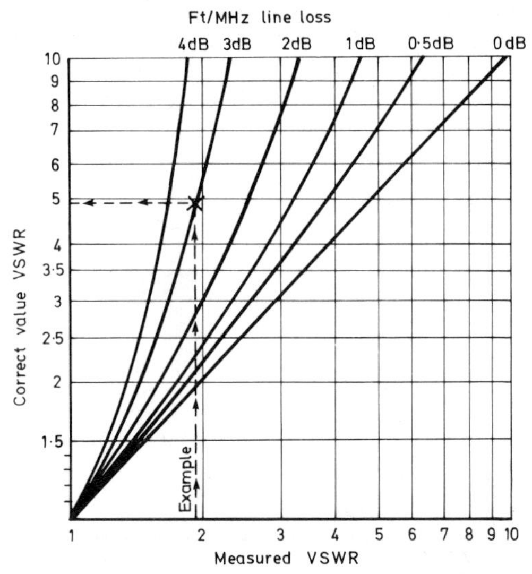

Fig. 5.3. Corrected v measured VSWR taking line loss into account; example shown by broken line

make depending on cable loss. It was stressed earlier that old cable as well as cheap inferior grade cable can exhibit high losses. The useful life of coaxial cable is about three to seven years, depending on climatic conditions, after which the loss factor may have increased by several decibels.

Simple power check

One of the simplest devices for checking power at the antenna elements is a small fluorescent tube of the kind used to light small spaces e.g. cupboards and cabinets. These are about 250 mm

(10 in) long and are rated at 6 W. Touching one of these tubes against a fully energised antenna element at a voltage point will cause it to light up at virtually full brilliance using about 10 W minimum power from the transmitter. Sometimes the tubes need to be warmed slightly to get them to strike and this is easily done by rubbing vigorously with a dry piece of cloth. Another useful device is an r.f. 'sniffer', which consists of an unbiased npn transistor (BC108 or similar) with an LED connected in series with the collector. The circuit is very simple as shown in Fig. 5.4 and can

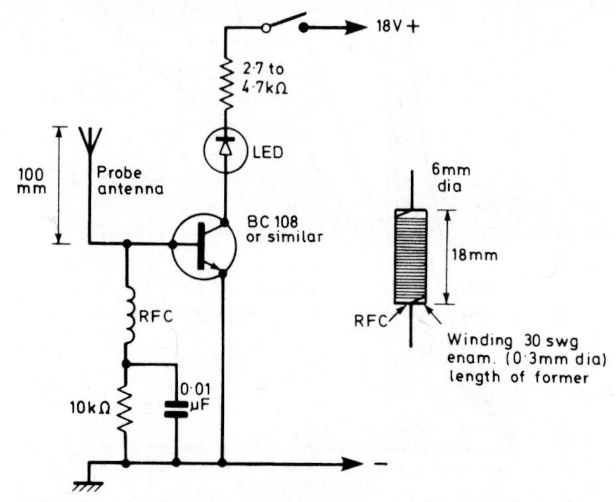

Fig. 5.4. A simple r.f. 'sniffer'; see text for construction and use

be housed complete with battery in a small metal box. The probe antenna need only be a piece of stiff copper wire about 100 mm (4 in) long for exploring the voltage maximum on driven and parasitic elements. Exploration should be carried out with the probe close to, but not touching the antenna element.

Effect of VSWR

Power lost in a transmission line is lowest when the line is terminated by a resistance the same value as the characteristic impedance of the line itself and becomes greater with an increase in the voltage and/or current standing wave ratio. This is due to the effective values of both current and voltage becoming larger as the standing wave ratio increases. The increase in effective current raises the resistive losses in the line conductors, and the increase in effective voltage increases the losses in the dielectric associated with the line. The increased loss caused by a VSWR greater than 1 or 2 may not be as serious as many think. If the VSWR is not more than 2 for instance the additional loss may not amount to more than about 0.5 dB even with long transmission lines. Fig. 5.5

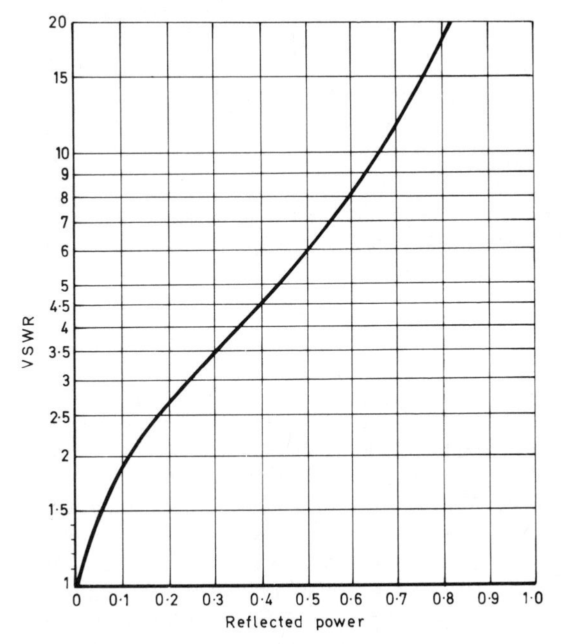

Fig. 5.5. Reflected power v VSWR readings

shows the percentage of returned (lost) power for various degrees of VSWR. However, the natural loss of the transmission line should be taken into account when checking VSWR as illustrated by Fig. 5.3. It should be noted also that when line loss is high and assuming a perfect match, the additional loss in decibels caused by a standing wave tends to be constant regardless of the line loss itself. The reason for this is that the amount of power reflected from the load is reduced anyway because, with high line loss, little power actually reaches the load in the first instance. As an example, if the line loss is 6 dB then only 25% of the power fed into the line from the transmitter will reach the load. If the VSWR due to mismatch at the load were 4 to 1, then 36% of the power reaching it would be reflected. As only 25% of the original power has actually arrived at the load the real reflected power will be $0.25 \times 0.36 = 0.09$ or 9%. This return power is further attenuated (by the line) by 6 dB, so that only $0.09 \times 0.25 = 0.025$, or a little over 2% actually arrives back at the transmitter output. This would result in a low VSWR reading at the transmitter end of the cable, in this case about 1.35 to 1. On the other hand with very low loss lines a high VSWR may increase power loss by a large amount although the total loss may still be fairly small by comparison with the power actually reaching the load. With a VSWR of 10 on a line having only 0.3 dB loss the additional loss would be about 1 dB.

Cable loss

Cable loss can be checked as mentioned earlier by measuring the power into a dummy load at the end of a cable and comparing this with the measured power going into the line at the transmitter end. If we assume 10 W into the cable and 9 W coming out into the load, then the loss in decibels is $10 \log_{10} P_2/P_1$ where P_1 is the power from the transmitter, i.e. 10 W, and P_2 the power into the load at the cable end which is 9 W. The loss is therefore $10 \log_{10} 9/10 = 0.457$ dB, or less than 0.5dB which can be considered as negligible. By the same token with a cable of given length and a known loss of say 3 dB, the power lost at the antenna would amount to half of what was fed into it at the transmitter end. We can work this backwards knowing that of 10 W originally fed to the cable only 5 W reaches the antenna. Therefore $10 \log_{10} 5/10 = -3$ dB. As the losses in cables and also due to mismatch etc. are relatively low in terms of decibels, Table 5.1 (power loss and gain

Table 5.1 LOSS OR GAIN IN DECIBELS FROM THE RATIO OF TWO POWERS

Power ratio (−dB)	*dB*	*Power ratio (+dB)*
1.000	0	1.000
0.977	0.1	1.023
0.995	0.2	1.047
0.933	0.3	1.072
0.912	0.4	1.096
0.891	0.5	1.122
0.871	0.6	1.148
0.831	0.8	1.202
0.794	1.0	1.259
0.707	1.5	1.413
0.631	2.0	1.585
0.562	2.5	1.778
0.501	3.0	1.995
0.398	4	2.512
0.316	5	3.162
0.251	6	3.981
0.199	7	5.012
0.158	8	6.310
0.125	9	7.943
0.100	10	10.000
0.079	11	12.59
0.063	12	15.58
0.050	13	19.95
0.039	14	25.12
0.031	15	31.62
0.025	16	39.81
0.019	17	50.12
0.015	18	63.10
0.012	19	79.43
0.010	20	100.00

e.g. power ratio of $\frac{P_2}{P_1} = \frac{10}{5} = 2$. Under +dB, the nearest figure is 1.995, which is equivalent to 3 dB.

in decibels) may be found useful. It is necessary to know only the ratio of the powers concerned. The same table can of course be used for determining the relative power gain between one antenna and another. For example, assume a dipole actually radiating 10 W of power is replaced by a beam antenna with 6 dB gain. What would be the effective radiated power of the beam? The power ratio for 6 dB is 3.981 or say 4, so the ERP from the beam would be 10 × 4 or 40 W.

The power loss in coaxial cables can be estimated by the graph in Fig. 5.6 by measuring the VSWR with a short circuit at the end of the cable. The example in the broken line shows that with a VSWR of 3 to 1 the cable loss whatever its length would be 3 dB.

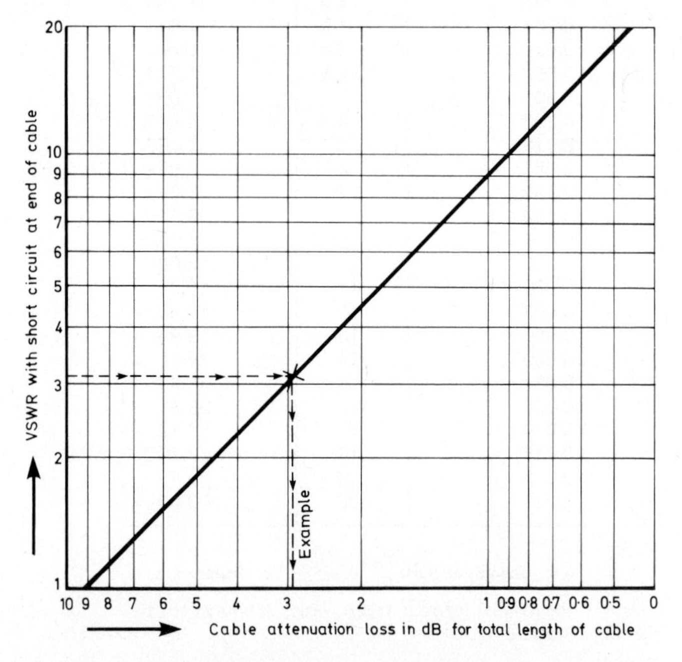

Fig. 5.6. Finding cable attenuation loss in decibels for a given length of cable by reading VSWR with short circuit at end of cable. Example: VSWR = 3. Cable loss = 3 dB as per broken line – see text regarding application of this test

This test should be made only if the cable is relatively long, as very high VSWR could result in damage to transistor output stages, although many modern transmitter/receivers for VHF and UHF have a built-in overload protection system that will immediately reduce or cut off power when a high VSWR is present.

Siting an antenna

Wave propagation at VHF and UHF is largely line of sight, as explained in Chapter 1, although atmospheric and other conditions do from time to time permit operation over long distances. The height of the antenna, for both transmitting and receiving, is therefore the most important factor in achieving long distance operation, with, secondly, clear surroundings. Ideally any VHF/UHF antenna should be so high that surrounding buildings and trees etc. are not in the line of radiation which is normally parallel to ground. Attenuation due to brick walls, metal framed buildings and trees is very high, higher than most realise. Experimental observations made on the effects of buildings have shown the radiation fields to have a median value about 25 dB below the expected over a clear path in the frequency range 40 to 450 MHz. Values have also been assessed for loss to be expected for transmission through trees when the antenna is lower than the tree tops, although in this case the loss depends on polarisation. At about 30 MHz the loss may be 3 dB for vertical polarisation and negligible for horizontal polarisation, but rising to 10 dB at over 100 MHz for vertical polarisation and 3 dB for horizontal. Brick walls offer some 6 dB or more attenuation at VHF and UHF, although this will vary depending on the dryness. It does, however, signify that the attenuation of radiation from indoor antennas could be relatively high, especially when the surrounding walls and roof are wet. Investigation has also been carried out with regard to the effect of terrain and buildings on the polarisation of received signals along the path of the wave. Diffraction and reflection from the ground, buildings and other conducting objects may cause the polarisation to become twisted. Some tests carried out by the author revealed that vertically polarised signals from stations at ranges of 30 to 40 miles, or more, can become orientated toward horizontal to such an extent that an increase in signal by 3 to 6 dB could be obtained when the receiving antenna was slanted to some angle between vertical and horizontal. Other factors that can greatly influence propagation are of course the phase of direct and reflected signals on arrival at the antenna, and these may vary very considerably with frequency, as illustrated in Fig. 1.5. The combined effects of diffraction, attenuation and phase arrival are most noticeable with signals from mobile transmitters, resulting in the characteristic 'flutter', particularly in built-up areas. Hence the importance of having an efficient and well placed antenna on the vehicle (see Chapter 2, mobile and portable operation).

Antenna gain

The gain of an antenna is the most difficult of all performance parameters that can be measured unless special facilities and equipment are available. As mentioned at the beginning of this chapter, something approaching free space conditions must be used if measurements are to be meaningful. The measuring equipment also must be very accurate, so conventional S meters calibrated in very doubtful decibels are quite useless, except for determining whether or not some increase or decrease in signal level has been obtained when one antenna is substituted for another. A further restriction is that the 'illumination' from the source antenna must be a plane wave over the effective aperture or capture area presented by the antenna being tested. Therefore the radiation source antenna and the one being tested must be a large number of wavelengths apart. Since 'free space' cannot be simulated the environment itself may affect measurement by reflections from the ground and other conducting obstacles in the vicinity of both the source antenna and the antenna being tested. Unless the aforementioned problems can be overcome, it is virtually wasted effort to attempt gain measurement with antennas mounted in normal surroundings. Much the same applies to plotting radiation patterns, for the true patterns of an antenna can be greatly distorted by reflection from other conductors (especially if these are resonant at the frequency of operation) and by ground reflection and reflection of signals from large buildings etc.; for this purpose also the measuring equipment must be very accurate.

Measurements with model antennas

The problems associated with the measurement of antenna performance have already been greatly emphasised. If one is to attempt this kind of work with a view to obtaining meaningful and reasonably accurate results, then not only is special equipment needed but also a special environment, this being the most difficult requirement, particularly if tests are to be carried out with full size antennas even at frequencies as high as 145 MHz. The author has been involved, over a large number of years, with establishing the performance of antennas for amateur and other frequency bands ranging from 1.8 down to 430 MHz, and like many professional workers, uses scale models of antennas operating in the UHF region and even at centimetre wavelengths. For purely amateur purposes a model antenna performance measuring system is not too difficult to set up, although if useful information is to be acquired, the equipment and procedures must be such as to leave little room for guesswork. References to two systems used by the author will be found at the end of the chapter, and each contains sufficient detail for a suitable set up and equipment for obtaining radiation patterns.

A model antenna test system

The equipment at present used by the author operates over a frequency band 650 to 1200 MHz and includes a separate 300 MHz transmitter used solely for impedance matching and phasing checks on antennas scaled to that frequency from lower frequencies. Virtually all work in connection with checking radiation patterns is carried out with the equipment shown in Fig. 5.7 and a special variable frequency transmitter (650 to 1200 MHz) incorporating its own broad band antenna and parabolic reflector.

At these frequencies the distance between the transmitting antenna and that being tested (in receive mode) and also their respective heights above ground, can be a reasonably large number of wavelengths, even though the area in which the system is used is fairly small. Care must be taken of course to eliminate reflection from conducting obstacles in the vicinity, preferably by removing such obstacles altogether. The equipment shown in the photograph consists of a 250 mm (10 in) diameter long-persistence CRT

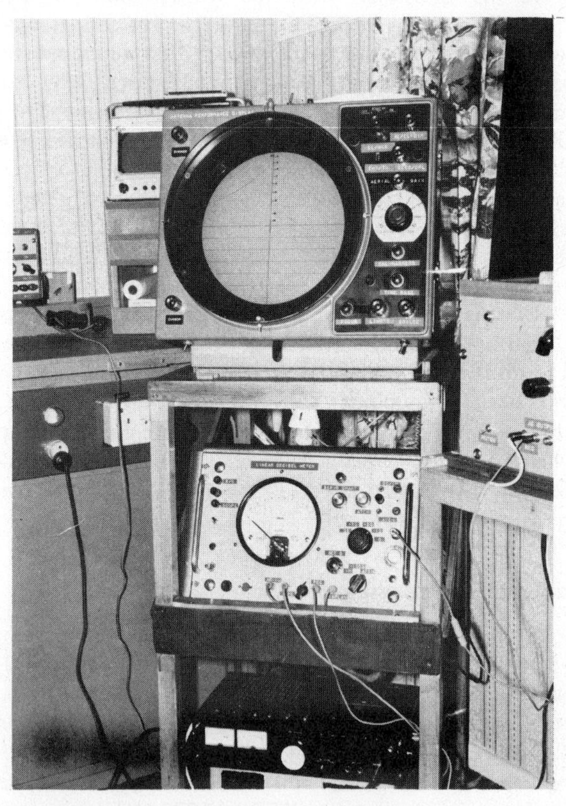

*Fig. 5.7. The **polar pattern indicator (PPI)** display and measuring equipment used by the author in antenna performance measurements*

display on which radiation patterns are displayed continuously as the time base rotates in synchronism with the antenna being tested. The measure of changes in amplitude is obtained by converting the varying signal from the antenna being tested into a variable width CRT brightening pulse. However, provision is made for gain measurement, usually against a reference dipole, with the aid of a linear reading decibel meter (below the PPI (polar pattern indicator) display in the photograph) which also incorporates provision for converting the varying antenna signals into a correspondingly varying amplitude low frequency (2000 Hz) waveform so that radiation patterns seen on the PPI display can be recorded on magnetic tape and recalled for display at any time. Calibration markers in log or linear dB scale and a 'roving' marker can also be displayed on the PPI as well as an electronically generated but very accurate 'reference' dipole pattern, or omni-

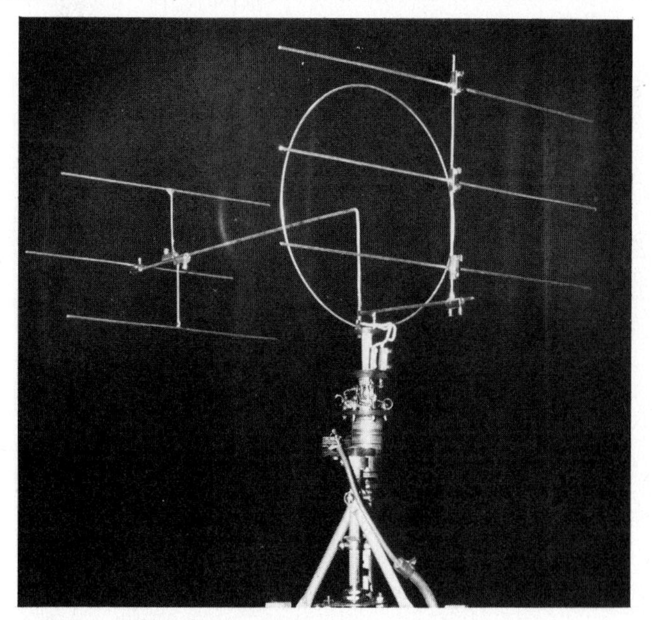

Fig. 5.8. Prototype model antenna. The slip ring coupler described in the text can be seen just beneath the antenna

directional (full circle) pattern, either of which can be set to any amplitude. These can also be superimposed on test antenna patterns. The display is equipped with illuminated graticules and maps and various colour filters for photographic purposes.

The model antenna rotating system is synchronised to the PPI by Selsyn motor drive, and each turns at about 1 revolution per 3 seconds. Antennas being tested are directly coupled to a rotating detector circuit which provides the d.c. signal via slip rings to the PPI signal-to-pulse converting circuitry and/or the dB meter etc. Most testing is carried out at 650 MHz as the antennas are relatively small, and because reasonably close matching to the detector is possible, and this is important if gain is to be checked with fair accuracy. The equipment readout is to within ±0.5 dB on amplitude, to within ±$^1/_3$ of a degree on pattern formation, and ±1 degree on angles associated with the patterns, e.g. establishing the −3 dB points for beam width.

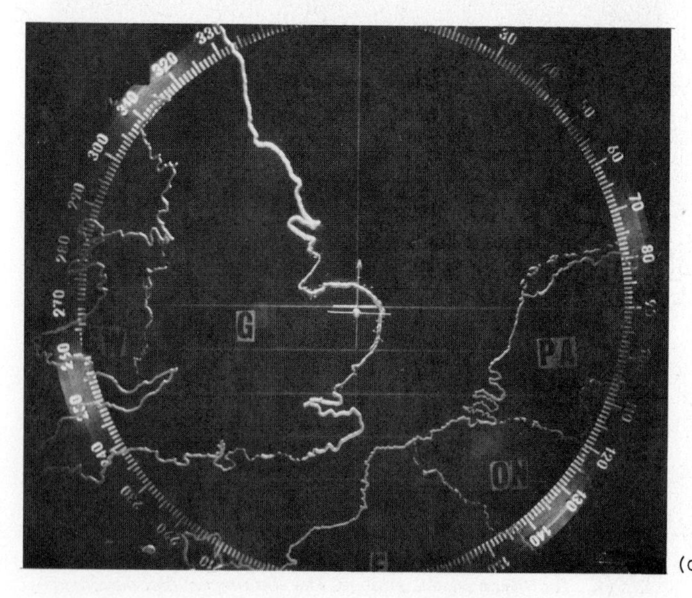

Fig. 5.9. (a) A map overlay used with the PPI display in Figure 5.7

The model antennas are usually made from copper or brass with elements scaled approximately to thickness as well as length etc. High grade low loss miniature coaxial cable is sometimes used to couple antenna to detector, although normally the feed point of a test antenna is coupled directly to the detector input. In some cases of course matching transformation may be necessary. Fig. 5.8 shows an experimental model antenna. The slip ring signal coupling can be seen just below the centre of the antenna.

A few of the facilities of the polar plotting display are illustrated by the photographs in Figs. 5.9; (a) is a typical map outline upon which polar patterns can be superimposed to give a more visual idea of the coverage of a particular antenna. Maps such as this can be drawn to any scale. The other photograph (b) shows the pattern from the electronic dipole which is a perfect cosine. It is generated by light projected through two pieces of polaroid material, one

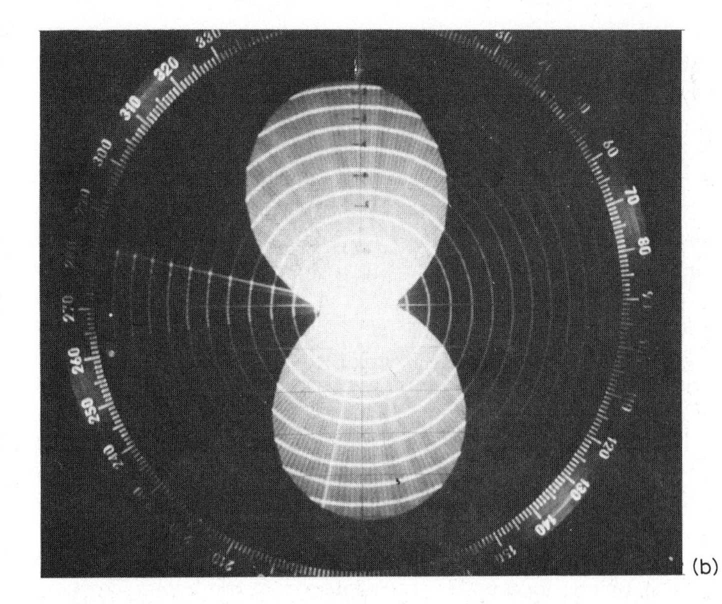

(b)

Fig. 5.9. (b) The 'electronic dipole' pattern and calibration marker rings (see text)

being fixed and the other rotating at the same speed as the PPI time trace. The light reaches a photocell, and, as the illumination intensity varies, the voltage output from the cell varies proportionally. This photograph also shows the calibration markers which have been set to a linear voltage scale.

Three examples of polar patterns obtained from model antennas are shown in Fig. 5.10. The radiation pattern (a) is from a 7-element ZL Special (described in Chapter 3) whilst (b) is from a 2-element ZL in vertical mode. The pattern (c) is from a typical Yagi type array and clearly shows the rear and side lobes usual with this type of antenna. Two of the photographs have a map overlay, and the circle trace is from the roving marker, in this case indicating the −3 dB points on the patterns.

Development of antennas with models

Aside from obtaining performance information from antennas of known design, models can be used to develop new designs. Here we can take the development of an antenna step by step from a basic idea to final formation and performance. The requirement is a compact array with relatively high gain and the design stems from the use of a loop radiator which initially has some gain over a dipole, a fairly high feed impedance to begin with and a radiation pattern the same as a dipole, i.e. figure-of-eight or cosine. The derivation of loop radiators from a folded dipole was shown in Fig. 3.26. There is however, another way of regarding the loop radiator in that it consists of two dipoles spaced distance *S* apart as in Fig. 5.11(a) but with the ends folded to meet, as (b), and thus form a square. Provided that the initial spacing (c) is not less than about 0.1λ, an oblong shaped loop as (d) can be formed, still with the advantage of a small gain but having the same radiation pattern as a dipole and high initial impedance at the feed points. In both cases the originally 'free' dipole is driven by the other since they join together at the ends.

Fig. 5.12(a) shows the stages of formation beginning with the driven oblong loop which is insulated from the boom. The next step (b) is a large area reflector to achieve the highest possible forward gain which, from a prototype model, was about 7 to 8 dB.

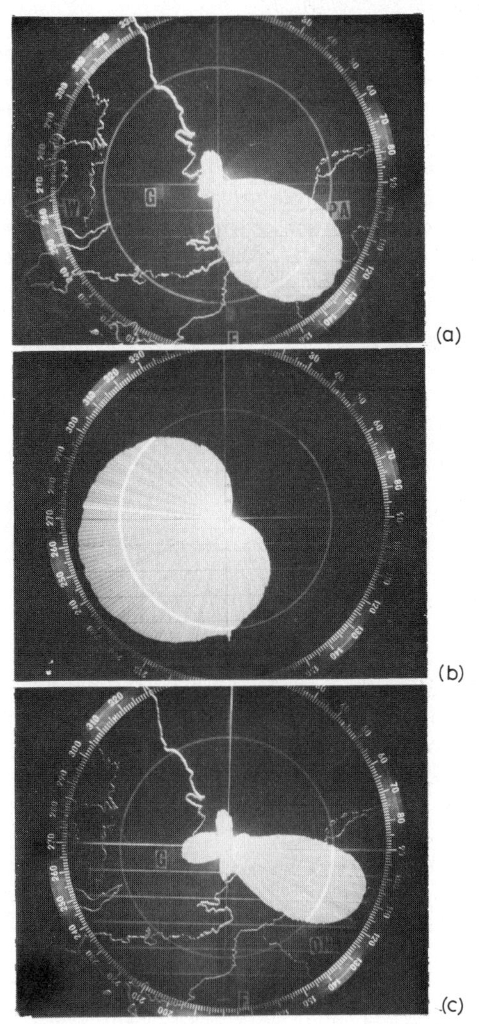

(a)

(b)

(c)

Fig. 5.10. Some typical polar patterns (a) of the ZL Special 7-element antenna in horizontal mode. (b) The cardioid pattern obtained with the ZL 2-element in vertical mode. (c) Pattern from typical Yagi type array which shows side and rear lobes

This forms an efficient driving system to which parasitic directors can be added as in (c) to further increase the forward gain. The dimensions for operation at 145 MHz (or other bands) can be calculated from the figures given in terms of wavelength and using the free space wavelength. For 145 MHz for example, this is 300/145 = 2.07 m. The element dimensions are therefore found by 2.07 times the fraction of wavelength given. The length of the oblong loop would be 2.07 × 0.375 = 0.77 m = 770 mm and so on.

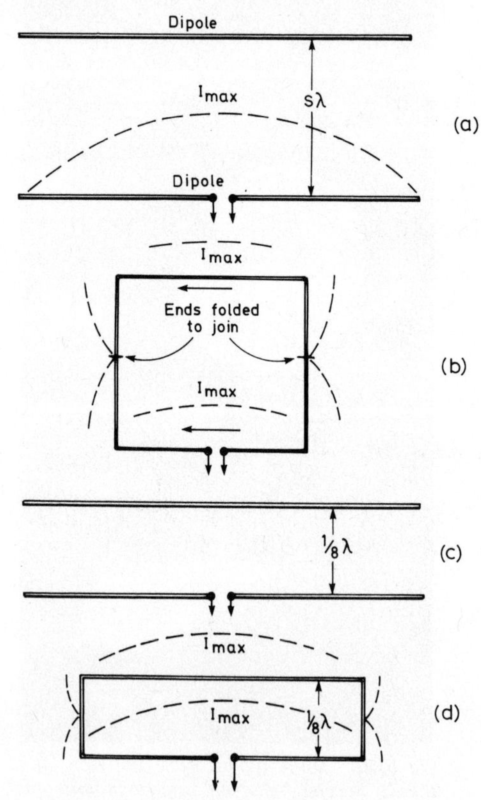

Fig. 5.11. Formation of loop radiators from a pair of spaced dipoles (a), (c) to form a square loop (b), and an oblong loop (d)

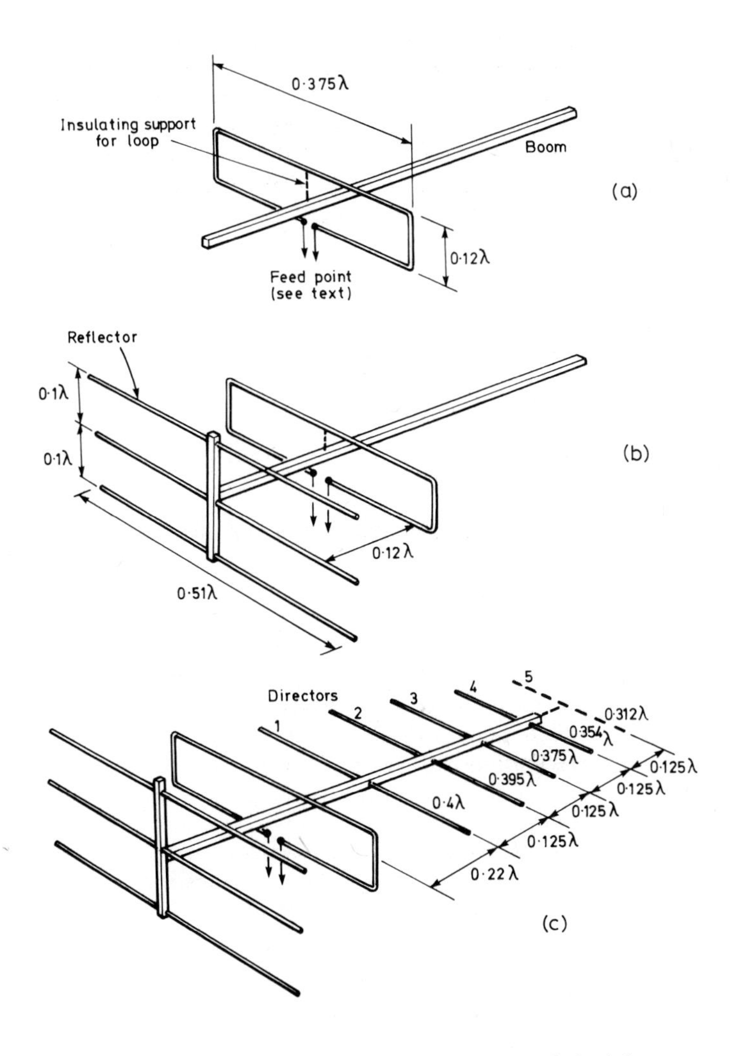

Fig. 5.12. Dimensions for a high gain beam antenna derived from performance and other data obtained with 650 MHz models (see text and polar patterns in Fig. 5.13)

The gain yield of this array should be in the region of 11 dB, but we can now see with the aid of the polar pattern display and with models how the final radiation pattern is developed. Fig. 5.13(a) shows the pattern of the loop radiator alone and with reference to the final gain obtained with the full array. The next stage (b) shows the effectiveness of the large three-element reflector providing a forward concentration of radiation with a gain factor in the region 7 to 8 dB. The final pattern (c) was obtained with four directors as in Fig. 5.12(c) and the measured gain was 11 dB. An additional director (shown dotted) produced a gain of about 12 dB ±0.5 dB. The total length of this array constructed for 145 MHz would be about 0.75 λ or 1.55 m (61 in). The feed impedance should be in the region of 50 Ω.

Field strength

There is little point in attempting to measure field strength from VHF aerials unless the measuring equipment is calibrated in either voltage or decibels, and even then readings obtained may well be seriously affected by reflection. A simple field strength meter is only really useful for establishing an increase or decrease in radiation, but even so it needs to be operated several wavelengths from the transmitting antenna. Field coverage of an antenna, or rather its effectiveness at range, is best estimated by continuous usage and comparison with another antenna. Signal reports from other stations can never be reliable as so called 'S' meters are not universally calibrated, receiver sensitivities vary considerably and so also may the antennas used by others. Two different stations at the same range but only a very short distance apart may well give entirely conflicting signal level reports. If it is a case of establishing whether a new antenna is better than the previous one, then carry out tests with one station only, first using the previous antenna as a reference and then replacing it with the new one in exactly the same position.

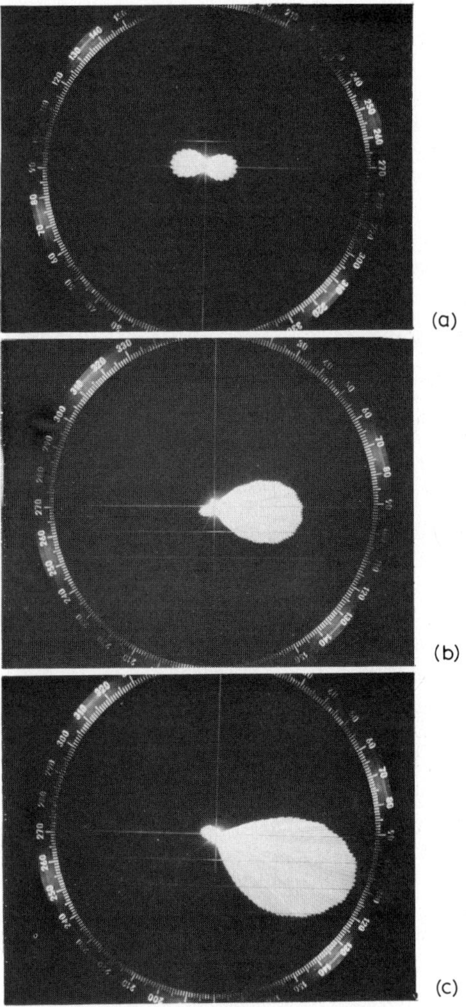

Fig. 5.13. How models show the performance at various stages of design of the antenna outlined in Fig. 5.12. (a) Cosine pattern with single loop radiator. (b) With addition of large area reflector. (c) Final pattern and gain with directors. Patterns are pro-rata with gain

Reading list

Subject	*References*
Antenna performance (general)	*ARRL Antenna Book,* American Radio League, available from Radio Society of Great Britain, 35 Doughty Street, London, WC1N 2AE.
As above but highly mathematical treatise	*Antennas*, J. D. Kraus, McGraw-Hill Book Co. Inc. (USA).
Antenna masts, design for masts, and wind loading on masts	*ARRL Antenna Book* (as above). *Ham Radio Magazine*, (USA), September, 1974. *Ham Radio Magazine*, (USA), August, 1974.
Antenna performance measurements with scale models	*Wireless World*, (UK), December, 1960 (F. C. Judd). *Practical Wireless,* (UK), January, 1978 (F. C. Judd).

Index